Franklin Lambert and Frits Berends

Einstein's Witches' Sabbath and the Early Solvay Councils

The Untold Story

Printed in France

© **2021, EDP Sciences**, 17 avenue du Hoggar, BP 112, Parc d'activités de Courtabœuf, 91944 Les Ulis Cedex A

This work is subject to copyright. All rights are reserved, whether the whole or part of the material is concerned, specifically the rights of translation, reprinting, re-use of illustrations, recitation, broad-casting, reproduction on microfilms or in other ways, and storage in data bank. Duplication of this publication or parts thereof is only permitted under the provisions of the French Copyright law of March 11, 1957. Violations fall under the prosecution act of the French Copyright law.

ISBN (print): 978-2-7598-2669-8 - ISBN (ebook): 978-2-7598-2670-4

Table of Contents

Foreword 1

THE FIRST PHYSICS COUNCIL 11

Chapter 1. A very unlikely "Council" 13
- 1.1 A quite surprising invitation 13
- 1.2 Novelty of the project 15
- 1.3 Ernest Solvay: industrial, scientific patron and investigator 17
- 1.4 The quantum theory 25
- 1.5 How Nernst discovered Einstein's genius 32
- 1.6 Nernst's great dilemma 34

Chapter 2. An unprecedented project 37
- 2.1 The idea of a Council 37
- 2.2 A providential man: Ernest Solvay 39
- 2.3 Nernst in action 41
- 2.4 A glimpse behind the scenes 46
- 2.5 Evolution of the project. 49
- 2.6 Ostwald's project for chemistry 58
- 2.7 Back to the Council 63
- 2.8 The first Council on Physics 68
- 2.9 Council results 78

UNEXPECTED CONSEQUENCES OF THE COUNCIL 93

Chapter 3. A game of musical chairs. 95
- 3.1 Impact on Einstein's career: from Prague to Zurich 95
- 3.2 The imbroglio of Lorentz's succession 99
- 3.3 Second impact of the Council on Einstein's career: from Zurich to Berlin 107

Chapter 4. Foundation of the International Solvay Institute for Physics 113
- 4.1 The Lorentz proposal 113
- 4.2 The Institute's Statutes. 116
- 4.3 Constitution of the International Scientific Committee (ISC). . . . 122
- 4.4 Birth of the International Institute for Physics. 123

Chapter 5. The second Physics Council. 139
5.1 Members and reports 139
5.2 Highlights of "Solvay II" 140
5.3 Echoes and consequences of Solvay II 146

Chapter 6. Foundation of the International Institute for Chemistry 149
6.1 Resuming contact with Ostwald 149
6.2 Intensification of Solvay's work 151
6.3 The chemists and their expectations..................... 152
6.4 Haller's good offices 154
6.5 Solvay's personal research 157
6.6 Back to ISIC 158
6.7 Impact of Ostwald's observations 160
6.8 Haller's master card 162
6.9 Culmination of a jubilee 163

Chapter 7. The Solvay subsidies. 167
7.1 Global situation 167
7.2 How to define the order of priorities? 171
7.3 Some notable successes 176
7.4 Last measures taken by ISC before the debacle 180

IMPACT OF THE GREAT WAR 183

Chapter 8. The Physics Institute survives the storm 185
8.1 First reactions to the invasion of Belgium 185
8.2 The Manifesto of the 93 188
8.3 A dead end conflict.................................. 193
8.4 Edition of the second Solvay volume: a thankless task 199
8.5 Solvay's actions and projects 201
8.6 Satisfactions, setbacks and hopes 206
8.7 Resumption of ISIP's activities......................... 212

Chapter 9. Epilogue: from "Solvay III" to "Solvay V" 219
9.1 Solvay III: Atoms and Electrons, April 1–6, 1921 219
9.2 Solvay IV: Electric Conductivity in Metals and Related Problems, April 24–28, 1924 223
9.3 Solvay V: Electrons and Photons, October 24–29, 1927. 226
9.4 Some final thoughts 238

ANNEXES 243

Annex 1. List of 52 Nobel laureates who took part in one (or in several) Solvay Councils between 1911 and 1933, or who benefitted from a Solvay research subsidy........... 245

Annex 2. Archival sources relating to the works of Ernest Solvay 247

Annex 3. Solvay's "Gravito-Materialitic" program 249

Annex 4. The Black-Body Problem 253
Annex 5. Planck's "missed" Nobel Prize 255
Annex 6. The second Moroccan crisis and the Caillaux affair 257
Annex 7. Royal patronage 259
Annex 8. Essential points in the Rutherford-Thomson
confrontation 261

Bibliography 265
Acknowledgments 271
Notes 273
Index 305

Foreword

The International Institutes of Physics and Chemistry, founded by Ernest Solvay are a unique concept in the world of science. These Institutes, originally separated into ISIP (International Solvay Institute for physics or International Physics Institute) and ISIC (International Solvay Institute for Chemistry or International Chemistry Institute), were created in 1912 and 1913[1]. They gave birth to the famous Solvay Councils, which were organized periodically for more than a century.

At the time of the Institutes' creation, international Congresses in physics and in chemistry were very rare, unlike the situation that we know today. The first Congress in physics took place in Paris in 1900[2]. It wasn't followed up, as physicists didn't feel the need for an international association. This state of affairs would remain unchanged until after the World War II[3]. Hence, we can say that the early Solvay Councils were unique, providing regular encounters between physicists of all nationalities

Each Council was convened for a definite purpose. It would address current questions in physics or chemistry, identifying the most urgent problems and opening avenues to their solution.

These specialized conferences were attended by a limited number of participants (less than thirty during the first years) invited on account of their expertise. Each question on the meeting's agenda was thoroughly discussed after an introductory report, presented by an especially competent invitee. The Council's reports were distributed among the members, preferably before the meeting. The discussions, recorded by the secretaries, were published afterwards together with the Council's reports.

The proceedings of the early Councils are a heritage of exceptional value. They bear testimony of the birth of modern physics and chemistry, providing a vivid image of major developments, such as the advent of the quantum era, the first steps in what is known as the physics of condensed matter, and the progressive disclosure of the fascinating architecture of atoms, with the development of atomic and nuclear physics[4]).

These celebrated Councils took place on a regular basis. They are still organized today, thanks to the active and generous support of five generations in the Solvay family.

A merger of the initial Solvay Institutes – ISIC and ISIP – was decided in 1970. They were transformed into an Association: the "International Institutes for Physics and Chemistry, founded by Ernest Solvay (IIPCS)". With the organization of periodic Councils, the IIPCS perpetuate Ernest Solvay's historic aim: to extend and deepen fundamental knowledge in physics and in chemistry by confronting the views of the most prominent experts in these fields.

The first Council's mode of operation became a recipe for success. Each Council was organized on the advice of an international scientific committee; it addressed a definite program and was attended by a list of well-chosen invitees.

The importance of the results of these Solvay meetings was recognized from the start, thanks to the impact of the first Council in physics. An exceptional standard of excellence has been maintained since then.

The Solvay committees for physics and for chemistry are at the heart of the IIPCS. Their periodic renewal guarantees a permanent link between each Council and the latest advancements in the corresponding field. This explains the presence at each Solvay meeting of several Nobel prize winners, and the attendance of the conference by future Nobel laureates[5].

It seems appropriate to note a parallel between the paths of Alfred Nobel (1833–1896) and of Ernest Solvay (1838–1922). Both men acquired a leading position in applied chemistry and left an indelible mark by putting their fortune at the service of humanity. Nobel's action was "downstream": he decided to award Prizes for discoveries in physics and in chemistry that should bring a lasting benefit to mankind. Solvay acted "upstream": he created international Institutes to encourage curiosity-driven research by means of subsidies and Councils. The task of the Councils was to address the most challenging problems, to detect the most promising directions for further investigation, and to provoke a better understanding of natural phenomena.

Solvay's action has sometimes been perceived[6] as the sign of a "more advanced internationalism" than that of Nobel. Indeed, by bringing scientists together from all over the world, the Solvay Councils encouraged the exchange of ideas about the latest problems in physics and in chemistry. They hereby generated new discoveries and paved the way to collective international research.

Singular aspects of the 1911 Council

It all started with a curious and most improbable event: the convening by Solvay of "*An international scientific Council to elucidate some current questions regarding the kinetic and molecular theories*".

The meeting was immortalized by a photo taken at Brussels' Hotel Métropole. The picture has travelled around the world. It shows a gathering of scientific celebrities:

Marie Curie, Henri Poincaré, Hendrik Antoon Lorentz, Max Planck, Albert Einstein, Jean Perrin, Paul Langevin, Marcel Brillouin, Walther Nernst,

Wilhelm Wien, Ernest Rutherford, Heike Kamerlingh Onnes, Arnold Sommerfeld, Martin Knudsen... and Ernest Solvay[7]. One also notices the presence of three Belgian scientists close to Solvay, Robert Goldschmidt, Edouard Herzen and Georges Hostelet, who had been asked to organize this *"unprecedented manifestation of higher science*[8]*"*.

The lasting significance of the first Council in physics was acknowledged in 2015 by the European Physical Society: Hotel Métropole[9] got the prestigious label of *"site of historical importance for the development of physics"*.

Yet, as we read the first reports[10] of this legendary meeting, we are struck by the fact that its origin has never been the subject of a thorough investigation. This lack of attention is all the more surprising as the convening of a "scientific Council" by an industrialist should have been regarded as an unusual, if not singular, operation.

Several authors took the trouble to examine the Council's proceedings. Some focused on the scientific context of the meeting[11]. Others emphasized its role with respect to the history of international conferences... But the Council's puzzling origin, remained largely undiscussed: no one seems to have taken notice of its strangeness.

As a contrast, some of the participants to the Brussels "summit" were surprised by its singular aspect. One reaction was that of Frederick Lindemann, a British physicist working with Nernst in Berlin, who felt prompted to publish a report in a German newspaper[12], under the title: *"A most curious Congress"*.

It is also known that Einstein complained in a letter to his friend Michele Besso[13] of having to interrupt his work to attend a *"witches' sabbath"*.

In fact, Solvay's convening of a "scientific Council" raises a number of questions:

— Why was the first international conference on quanta organized in a country where nobody cared about their existence?
— Why was this summit launched by a patron of science, and not by a professional physicist?
— Why was Solvay concerned by a problem which only worried a handful of scientists: the accumulation of experimental evidence against classical kinetic and molecular theories?

Other relevant questions are about Solvay's foundation of an International Institute of Physics, one of the consequences of the Council's success:

— Why did Solvay grant priority to physics after having said that he would support the creation of an International Institute for Chemistry?
— Why did this passionate investigator, working on his own theory of the Universe, accept to found an Institute for Physics in which he would have *"nothing to see nor to say"*?
— And why did Solvay fail in his attempt to create at the same time two similar Institutes: one for physics (ISIP) and one for chemistry (ISIC)?

It is our aim to answer these questions by retracing Ernest Solvay's most spectacular realizations in science: his convening of the first Council on physics (1911), and his subsequent foundation of ISIP (1912), followed by that of ISIC (1913).

Let us recall straight away that the first Solvay Council played a crucial role in the development of quantum theory. Launched 10 years after Planck's formulation of his quantum theory, it showed that the "extravagant" quantum hypothesis could be applied with success to a problem (the nonclassical behaviour of specific heats) that had nothing to do with the problem for which it had been introduced[14] (the black-body problem). Thus, by publicizing the successful exportation of Planck's hypothesis from the field of radiation to the field of matter, the Council opened the way to its general acceptance.

A first step in this direction was taken with the consensus that classical theories failed to account for the observed properties of thermal radiation and of the low temperature behaviour of matter. The Brussels meeting can therefore be seen as "drawing a watershed[15]" between two eras: the classical era which culminated with the formulation of the Maxwell–Lorentz electromagnetic theory of light, and the quantum era which originated with the works of Planck and Einstein.

With respect to methodology, the Council inaugurated a new working method.

Today, it is the method adopted in a "scientific workshop", a meeting of scientists engaged in the study of a common problem.

Let us now say a word about Solvay's first international foundation (ISIP). This Institute surprised the world by the novelty of its actions. Never before had physicists from all countries been given the opportunity to apply for support to an "international scientific committee". ISIP's program of research subsidies, an idea introduced by Solvay, and brought into life by Lorentz, made a huge impression.

Its success was immediate: six beneficiaries of a "Solvay grant" saw their work rewarded by a Nobel prize: Max (von) Laue (subsidy in 1912, Nobel prize in 1914); Charles G. Barkla (subsidy in 1912, Nobel prize in 1917); James Franck and Gustav Hertz (subsidy in 1913, Nobel prize in 1925); William L. Bragg (subsidy in 1914, Nobel prize in 1915); Johannes Stark (subsidy in 1914, Nobel prize in 1919).

In fact, it is worth recalling that ISIP's first move took place before the start of its official subsidy program. This action, the granting of financial support to a Russian laboratory in distress (Lebedew's laboratory in Moscow), was a first and spectacular illustration of what an International Institute for Physics could do.

Equally amazing is the observation that international celebrities, such as Lorentz, Marie Curie and Ernest Rutherford, didn't hesitate to spend part of their precious time to evaluate the 97 demands of subsidies that flowed into the Institute between June 1912 and July 1914. This disinterested commitment of top researchers is a moving testimony of their faith in the universal role of science and in its value as a contributor to peace, a faith that, as we know, would soon be annihilated by the First World War.

The story that we are going to tell is made of several layers. Each layer consists of apparently unrelated elements, but which turned out to be linked by unforeseen circumstances.

A major factor was the crisis that worried physicist at the beginning of the twentieth century. Another factor was Solvay's profound predilection for science, a passion that would lead him to accept the unrelated requests of two German chemists, Walther Nernst and Wilhelm Ostwald, who needed his support to achieve their personal goals.

We shall see that nothing would have been possible without Solvay's indefectible confidence in his power to build an alternative to generally accepted physical theories, but incapable in his eyes of satisfying the philosophical spirit. It will also become clear that no Council would have come into being, had there not been an unexpected "parallelism" between Solvay's very personal views and the worrying situation in physics: the growing disagreement between experimental facts and the predictions of classical theories.

A number of elements are known: they have been reported by major historians of science who focused their attention on the scientific background of the early Solvay Councils. Yet, several factors, in particular those related to Solvay's personal agenda, haven't been taken into account. Without these factors, it is hard to understand the reasons which pushed the industrialist to convene a scientific Council in June 1911, and to prolong its effects through the foundation of two international institutes.

It is to this difficulty that we have tried to respond by retracing Solvay's commitment in the light of unpublished archival documents[16].

We shall see that Solvay was absorbed by his own theories when he was approached by Nernst (June 1910) and by Ostwald (April 1911). The former asked him to send a letter to twenty prominent physicists, inviting them to attend a "*Konzil*" in Brussels. The latter approached him with an ambitious plan: that of creating an International Institute for Chemistry, in partnership with the International Association of Chemical Societies.

These totally unrelated requests benefitted from a happy circumstance. They were seen by Solvay as a way to fulfil a wish he had long cherished: to promote the increase of knowledge in physics and in chemistry by acting in two "contrasting" directions.

One of these directions aimed at serving a personal ambition. Solvay dreamed to have his scientific ideas validated by an authorized learned body. However, this personal program didn't prevent him from supporting professional researchers in their current scientific pursuits... Indeed, motivated by his second goal – to encourage the study of recent phenomena that might disprove generally admitted theories – he welcomed new experiments[17], in particular research on two challenging problems on which he had personal views: Brownian motion and radioactivity. Aware of the lack of Belgian experts, he felt ready to support researchers "from all over the world" in a spirit of objectivity and of scientific internationalism.

Solvay's endeavours in the first direction didn't produce the success he expected. In contrast, those which aimed at the other goal proved extremely fruitful: they led to the creation of two International Institutes, ISIP and ISIC, and to the celebrated Solvay Councils, which would perpetuate the founder's wishes far beyond his expectations[18].

ISIP was created in a few months, thanks to the wisdom and to the dedication of Lorentz, a theoretician of great repute and a man of tact whose ideas agreed perfectly with Solvay's official goal (just as Solvay, Lorentz regarded science as a way of favouring human welfare; as Solvay, he had spent his early research in "splendid isolation").

The creation of ISIC was quite another matter. It proved extremely complicated, due to an incompatibility between Solvay's views and the priorities of the International Association of Chemical Societies (IACS). The disagreement was such that Solvay saw no other way than to abandon some conditions he had set to his intervention. In return, he had the joy in 1913 to see his industrial jubilee heightened by the holding of an international summit of chemists (the second general meeting of the IACS). He also had the satisfaction of witnessing the success of a plan which had cost him so many efforts: the foundation of ISIC.

It would have been unwise to turn a blind eye on Solvay's obstinacy to build a "*Gravito-Materialitic*" theory, basis to all his designs[19]. There is no doubt that his stubbornness in this activity, comparable to the one he displayed in the construction of his industrial empire, was the element which led to success in his other undertakings: the realization of Nernst's "Konzil" in 1911, and above all the prolongation of the "Conseil de physique" by the creation of ISIP.

Hence, it is fair to say that Solvay's unrealistic project, but one that should be placed in the context of the time, has been the "key" to the achievements that we are going to examine.

On the other hand, we feel that by ignoring Solvay's ambition we would do insult to the memory of a great visionary, a man who once confessed that "*industry had only been for him a means of acquiring the independence needed to do science*[20]".

Any reader of Solvay's "preliminary study[21]", distributed to the Council members, and of the many Notes[22] of his scientific diary for the years 1910–1912, can only be impressed by the efforts made by this self-taught man in order to complete his "great work". Solvay's determination to achieve this goal clearly appears in his opening and closing speeches of the meeting. It explains his initial "ambiguous" position with respect to the newly founded Institute for physics, and gives a particular sparkle to his later decision

Foreword 7

not to interfere in any way with its management. By accepting Lorentz's proposition to entrust the scientific direction of ISIP to an International Committee, whose composition would evolve over time, Solvay showed admirable clairvoyance.

About the book

It seemed appropriate to evoke the scientific developments that led to the first Council in physics and to the subsequent foundation of ISIP. These reminders, limited to the facts related to our story, are necessary to understand the motivations of Solvay's scientific partners. However, they may be discarded by readers who aren't familiar with the birth of modern physics... Indeed, we believe that they shouldn't prevent these readers from following the course of events in this human adventure, made of passion, ambition, and conflicts.

We devote the first part of the book to the first Council in physics: its origin, its convening, proceedings and main results. This part, entitled "The first Council on Physics", comprises two chapters, devoted to the following points:
- Singularity of the invitation of June 1911 (1.1 and 1.2).
- Personality of Ernest Solvay; reasons which led to his convening of a "Scientific Council" at Walther Nernst's request (1.3, 2.2 and 2.3).
- Elements of quantum theory; Nernst's initiative (1.4, 1.5 and 1.6).
- The Nernst project, evolution and preparation (2.1, 2.4, 2.5 and 2.7).
- The Council of 1911. Opening speeches, council meetings, minutes of the conference, impressions of some members, closing speeches (2.8).
- Results of the meeting, general impression and positive elements, press reactions, Solvay's attitude after the Council (2.9).

NB: Section 2.6 is devoted to Ostwald's proposal for chemistry.

The second part of the book, "Unexpected consequences of the meeting", starts with a chapter on the impact of the Council on the development of Einstein's and Lorentz's professional careers. A second chapter deals with the foundation of the International Solvay Institute for Physics. We recall the fortunate coincidence between Solvay's wish to call on Lorentz's services, and the latter's intention to give up his chair at the University of Leiden.

We devote two chapters to Solvay's creation of ISIC, in partnership with the International Association of Chemical Societies (IACS). This digression with respect to our main purpose seemed justified on account of two facts:
- Ostwald submitted a plan for an International Institute for Chemistry at the time of the opening of the Council.
- Solvay welcomed Ostwald's plan insofar as it allowed him to found two similar Institutes, one for physics and one for chemistry. However, the idea proved illusory, and the creation of ISIC would demand efforts and renunciations on Solvay's part which surpassed everything he had agreed to for the creation of ISIP.

To summarize, the second part of the book comprises five chapters (with numbers ranging from 3 to 7) on the following subjects:
- A game of musical chairs: impact of the Council on the careers of Einstein and Lorentz (3.1, 3.2 and 3.3).
- Foundation of the International Solvay Institute for Physics (4.1, 4.2, 4.3 and 4.4).
- The second Council in Physics (5.1, 5.2 and 5.3).
- Foundation of the International Solvay Institute for Chemistry (from 6.1 to 6.9).
- The "Solvay subsidies" (7.1, 7.2, 7.3 and 7.4).

In the third part of the book, "Impact of the Great War", we report elements related to the first world conflict, such as the collapse of international scientific relations, and the "miraculous" survival of ISIP. We also comment on an unexpected benefit for ISIC: the dissolution of the IACS. It enabled Solvay to fulfil his wish to create an independent Institute for Chemistry, on the model of ISIP.

This third part contains two chapters. The first one (chapter 8) is devoted to the following points:
- First reactions to the invasion of Belgium (8.1).
- The *Manifesto of 93 German intellectuals*; reactions to this pamphlet (8.2).
- The behaviour of physicists from different countries who had been invited to pre-war Councils, or who had benefitted from a Solvay-subsidy; actions undertaken by Lorentz to ease tension between physicists from both camps; Nernst's secret mission; Sommerfeld's surprising attitude (8.3).
- The thorny question of the proceedings of the second Solvay Council (8.4).
- Solvay's project for the construction of a post-war society with the help of ISIP and ISIC; reactions of Lorentz, Haller and Guye; Berlin preventing Solvay's visit to Switzerland in July 1918 (8.5 and 8.6).
- Reorganization of ISIP before resuming its activities; setting up of a new scientific committee (8.7).

Chapter 9 is an epilogue. We present a short overview of the post-War Councils that took place under the presidency of Lorentz: Solvay III (9.1), Solvay IV (9.2), Solvay V (9.3). We recall the measures taken in 1926 to bring an end to the breakdown of the international dialogue. We conclude our story with some final thoughts (9.4).

One of our goals has been to describe the aspects of a physicist's life at the start of the twentieth century. We discuss the uses that were current at the time in the academic world (some practices still exist today).

Throughout the narration, we sought to get as close as possible to the actors of this adventure in order to enter their way of thinking. We therefore chose to frequently give them the floor, a choice which requires a great number of citations.

Einstein's letters, from which we reproduce numerous excerpts, can be found in *The Collected Papers of Albert Einstein* (*CPAE*), vol. 5, *The Swiss Years, Correspondence 1902–1914*, edited by Caltech, published in 1995 by Princeton University Press, and available today on *Digital Einstein Papers*.

As to the copyright of Einstein's letters, they belong to the Hebrew University of Jerusalem.

Other citations have been extracted from letters written in French, German or Dutch.

Our willingness to pull ISIP's subsidy program from oblivion (it was its main activity during the years 1912–1914), prompted us to recall the unprecedented measures that were taken by Lorentz to set up an *objective system* for the evaluation of scientific projects, a "world premiere" at the time. Since then this question has only grown for those responsible for the financing of fundamental research in physics and in chemistry. We hope that this reminder will give an additional value to the story we have chosen to tell. Readers who aren't interested in technical details can omit sections 7.1 and 7.2.

THE FIRST
PHYSICS COUNCIL

Chapter 1
A very unlikely "Council"

1.1 A quite surprising invitation

On 15 June 1911, about twenty physicists received the following confidential message: *"Invitation to an international scientific Council to elucidate some current questions regarding the molecular and kinetic theories"*.

> *Dear and very honoured Sir,*
>
> *By all appearances we are at the moment in the middle of a new evolution of the principles on which the classical molecular and kinetic theory of matter is based.*
>
> *On the one hand, this theory, in its well-reasoned development, leads to a radiation formula that disagrees with all the experimental results; on the other hand, from the same theory predictions arise on the behaviour of specific heats (law governing the variation of the specific heat of polyatomic gases with temperature, validity of the rule of Dulong and Petit at low temperatures) which are also refuted by numerous measurements.*
>
> *As has been shown in particular by Messrs. Planck and Einstein, these contradictions disappear when certain limits are imposed to the motion of electrons and atoms in the event of oscillations around a rest position (assumption of energy degrees). Yet, this idea, in turn, deviates so considerably from the equations of motion of material points that have been used so far, that its acceptance would necessarily and unquestionably require a major reform of our fundamental theories.*
>
> *The undersigned, although alien to special questions of this kind, but animated with a sincere enthusiasm for all problems which by their study broaden and develop our knowledge of nature, believes that a written and verbal exchange between researchers more or less directly concerned with these problems, could – if not lead to a final decision – at least clear the way to their solution.*
>
> *A big step in the development of atomistic theories would already be taken if we could clearly establish which of our molecular and kinetic*

interpretations agree with the results of observation, and which, on the contrary, should undergo a complete transformation.

With this goal in mind, the undersigned proposes you to participate in a "scientific Council" to be held in Brussels from Sunday 29 October to Saturday 4 November 1911, bringing together in a restricted committee a few eminent professionals. It would be composed as follows:

Chairman: Mr. Lorentz (Holland);

Secretaries: Messrs. R. Goldschmidt (Belgium), M. de Broglie (France);

Members: Messrs. Jeans, Larmor, Lord Rayleigh, Rutherford, Schuster, J. J. Thomson (England); Nernst, Planck, Rubens, Sommerfeld, Warburg, W. Wien (Germany); Brillouin, Mme Curie, Langevin, Perrin, H. Poincaré (France); Einstein, Hasenöhrl (Austria); Kamerlingh Onnes, van der Waals (Holland); Knudsen (Denmark).

The subjects to be dealt with are:

1. *Deduction of Rayleigh's radiation formula.*
2. *Comparison of the theory of ideal gases with the results of experiment.*
3. *Application of kinetic theory to emulsions.*
4. *The kinetic theory of specific heat according to Clausius, Maxwell and Boltzmann.*
5. *The radiation formula and the theory of energy degrees (Quantenhypothese).*
6. *Specific heats and the theory of energy degrees.*
7. *Application of the theory of energy degrees to a series of physical problems.*
8. *Application of the theory of energy degrees to a series of physico-chemical and chemical problems.*

For each of these questions we shall ask an especially competent member to kindly write a preliminary report. These reports, written in French, in German or in English, will be printed and distributed to all members, if possible before the end of September. They will later be collected in a volume, together with a record of the discussions.

Being not a specialist in science, I won't be able to deal with the above subjects.

Yet, having made since long a general study of gravity with the view to drawing consequences on the constitution of matter and energy, I propose to provide a brief summary of this study at the opening of the meeting, considering that it could have an influence on some of the works.

To allow all invitees to participate, I offer everyone an indemnity of 1000 francs for travel expenses.

Possible questions and responses should be addressed to Prof. Dr. W. Nernst, Am Karlsbad 26a, Berlin W.35.

I hope I can count on your collaboration, and beg you to accept, very honoured Sir, the assurance of my highest consideration.

Ernest Solvay

What can we say about this invitation?

The convening of an International Congress of physicists in 1911 was by itself an exceptional event. The only precedent was the Paris Congress, which in 1900 brought more than 750 physicists together, representing 24 countries.

Another peculiarity was the fact that the Paris Congress had been organized by the French physical society, whereas the present invitation letters were signed by an individual, acting in his own name.

Of the 24 invited members, only 4, Nernst, Planck, Lorentz and Knudsen, had been previously informed. The others had serious reasons to be surprised.

— Was there really a crisis in physics? And if so, why was it a Belgian industrialist who alerted the scientific world?
— Why a Scientific Council in Brussels when Belgium is only represented by an obscure secretary?
— What was Nernst's role? Wasn't he the obvious author of the scientific argument used by Solvay to justify the convening of a Council?

The perplexity of most of the guests was increased by the *confidential* character of the invitation... Why this concern for secrecy?

The singularity of the operation provoked mistrust of renowned scientists, such as Lord Rayleigh, Sir J. J. Thomson and J. D. van der Waals, who declined the invitation under various pretexts.

Einstein, who was mentioned as one of the rapporteurs, had been informed by Nernst. He immediately understood his role. These are his words in a letter to Nernst[23] of June 20, 1911: "*The whole enterprise pleases me tremendously...*".

1.2 Novelty of the project

Solvay's initiative caused, as we have seen, considerable surprise. It still raises questions today. We shall try to address each of them by examining the facts which led to the following developments: the suspicion that physics was in crisis, the idea that the situation required the holding of a Council, and the extraordinary suggestion that this Council should be convened by Solvay.

Let us first take a look at the meetings' picture taken at Hotel Métropole (figure 1) on the second day[24] of the Council. A simple glance reveals a staging, apparently wanted by the art photographer Benjamin Couprie. Indeed, we notice that Nernst (first seated person from the left), Wien and Rutherford (sitting and standing in fourth position from the right) have their eyes fixed on the photographer... Conversely, Poincaré, Marie Curie and Perrin (seated at the right) seem totally absorbed by some reading or by a discussion.

FIGURE 1: Group photo, taken on October 31 at Brussels' Hotel Métropole. Service des archives et des bibliothèques de l'Université Libre de Bruxelles (S.a.b.ULB). Courtesy of the International Institutes for Physics and Chemistry founded by Ernest Solvay. Sitting from left to right: Nernst, Brillouin, Solvay, Lorentz, Warburg, Perrin, Wien, M. Curie, Poincaré. Standing from left to right: Goldschmidt*, Planck, Rubens, Sommerfeld, Lindemann*, M. de Broglie, Knudsen, Hasenöhrl, Hostelet*, Herzen*, Jeans, Rutherford, Kamerlingh Onnes, Einstein, Langevin (the names with a * are those of scientific secretaries or of Solvay's assistants).

Another obvious point is the fact that the photo was retouched in order to present Solvay, who wasn't present at the meeting, sitting at the end of the table between Marcel Brillouin (at his right) and Chairman Lorentz (at his left).

Brillouin, as we shall see, was Solvay's personal guest. This explains his prominent place on the picture. But the Frenchman looks worried. As he would later confess, he hadn't been sure to accept the invitation because he understood neither German, nor English...

In fact, it isn't easy today to measure the surprise caused by the need to discuss eight reports presented in three languages, and to capture on the spot the abundant remarks formulated in German, French or English[25]. This difficulty proves by itself the novelty of the meeting and the risks taken by Solvay. The challenge was such that it could have led to a failure of the operation...Indeed, we know that Lord Rayleigh used the "linguistic obstacle[26]" as a pretext to justify his absence.

Fortunately, the operation proved a success, thanks to the talent of Chairman Lorentz and to his perfect mastery of the three languages. His outstanding role in the conduct of the debates will be commented on in section 2.8.

Let's focus first on the signatory of the Council's invitation letters.

1.3 Ernest Solvay: industrial, scientific patron and investigator

Founder of the ammonia-based soda industry, Ernest Solvay (figure 2) was a genius inventor[27]. Having been told that he wasn't the discoverer of a reaction that might be used to produce soda (the reaction was known since 1811), he didn't give up... On the contrary, he concluded that it was a good idea since someone else had already thought of it, and that he would succeed where others had failed.

Indeed, relying on his family, Solvay managed by indestructible tenacity, and with the help of his brother Alfred, to build an industrial empire. His relentless efforts to make the process profitable were crowned with success to the point of eclipsing the existing manufacturing process, developed in 1787 by Nicolas Leblanc.

FIGURE 2: Ernest Solvay in 1895. Courtesy of the Solvay family.

Philanthropist and founder of research institutes, Ernest Solvay was much more than a patron of science, like a Carnegie or a Rockefeller. Driven by passion, this self-taught entrepreneur devoted himself to scientific investigation. Research became his main concern from 1885, when he asked his brother to ensure the management of the family business. Hence, we may place this extraordinary man in the line of Gaius Cilnius Maecenas, the Roman statesman who supported poets while practicing their art.

This is how Solvay defined himself[28] in 1893:

> "I am not fortunate enough to be a man of science: I didn't receive a classic education, the problems of the industry have consumed my time. But it is true that I have never ceased to pursue a scientific goal, because I love science and expect from it the progress of mankind."

As for his research program, he summarized it as follows[29]:

"I saw in the new paths of science three directions that I followed, three problems which in reality, in my eyes, only form one. First, a problem of general physics: the constitution of matter in space and time; second a problem of physiology: the mechanism of life, from its most humble manifestations up to the phenomena of human thought; and third a problem complementary to the first two: the evolution of the individual and of social groups."

In fact, Solvay's commitment to science was based on the idea that general welfare could only result from a unified development of physical, natural and social sciences, called to merge into a "universal science". Seeking to make progress in these three directions, which he believed he could link through an "energetic approach", the industrialist surrounded himself with professionals, urging them to assist him in the creation of specialized institutes. A first project aimed at the creation of an institute, designed to verify his own theories. Solvay formulated the idea in a note of April 1880, entitled[30]: *"Research base for my Institute"*. Abandoned at first, the project resurfaced during the 1890's. The industrialist then decided to go one step further, with the creation of a "City of Science" at Brussels' Park Leopold, in partnership with the municipality and with Brussels' Free University.

The project's first realization was the foundation, in 1894, of an Institute of Physiology. It was followed in 1902 with the foundation of an Institute of Sociology, considered as a natural complement to the previous Institute (the two buildings still exist today).

The Physiology Institute was headed by Professor Paul Héger, a brilliant physiologist who happened to be Solvay's personal physician. The building (figure 3) housed two separate Institutes. One was dedicated to university education. The other depended directly on the founder: its mission was to develop Solvay's views on the role of electricity in the phenomena of life.

FIGURE 3: The Solvay Physiology Institute, Parc Léopold (Brussels). Digithèque de l'ULB, digithèque des IIPCS. Courtesy of S.a.b.ULB.

The Institute of Sociology succeeded an earlier "Institute of Social Sciences", set up in 1894 and housed in Hotel Ravenstein. Solvay, a liberal (*i.e.*, conservative) senator didn't hesitate to entrust the Institute's direction to three socialists, H. Denis, G. De Greef, and E. Vandervelde, whose ideas were close to his own[31] – a decision which bears testimony of his freedom of mind. Unfortunately, differences of opinion soon emerged, putting an end to the experiment[32]. Solvay realized that it was better to deal with only one director. He founded a new institute in Park Leopold and entrusted its direction to Emile Waxweiler, an expert who had been a teacher to King Albert. Under Waxweiler's savvy leadership, the Institute of Sociology became a laboratory responsible for developing the founder's sociological ideas. The proximity of the two Institutes responded to Solvay's wish to illustrate the link which he sought to establish between the laws of physiology and those which govern human societies.

But what about Solvay's first problem? Why isn't there any trace in Park Leopold of an Institute of physics, where the industrialist's ideas on the constitution of matter could have been deepened and submitted to a check by experiment?

The most likely reason is that Solvay didn't find a person of confidence, capable of leading such an Institute.

Unlike the Netherlands, Belgium had no renowned physicists. In chemistry, there had been Jean Servais Stas, a scientist celebrated for his accurate determination of atomic weights. But this holder of a Davy Medal, had died in 1891.

We know that Solvay consulted Stas on his projects and scientific intuitions, but that he received no support[33]... We will come back to Stas' discouraging reactions, and to their impact on the industrialist's attitude towards leading experts (see section 2.9).

On the other hand, Solvay received a warm welcome from another, most successful chemist: Ostwald, holder of a Nobel Prize, and promoter in Germany of the "energy doctrine[34]".

Here is Ostwald's judgement[35] on the man he described as the "founder of social energetics":

> "*As is often the case with self-taught geniuses, Solvay's results were a mixture of enriching ideas and clumsy mistakes. Since it is above all the latter that emerge from a superficial reading, his attempts to interest other researchers in his ideas received only limited attention. Anxious to provide them with a place of development, he founded an Institute and endowed it richly, thanks to the millions he possessed. However, true to his principles, he did not content himself with responding to a personal passion: he created under the same conditions two distinct institutes, one for physiology, the other for commercial studies*[36]. *He took care to remunerate generously the directors of his institutes, allowing them to live with their families without having them to resort to another job. Determined to preserve their scientific independence, he guaranteed their employment by contract, whatever their attitude towards his views...*".

Solvay's passion for physics

According to Emile Tassel, Solvay's early scientific advisor, it was physics that constituted the basis of his research program.

The industrialist's need to discover the laws that govern the world originated from a curiosity that inhabited him since childhood. Speaking in 1897 to his sister Aurélie, Solvay said[37]: "*I must penetrate the problem of the Universe*[38]". His determination was just as strong in 1913 when he told Ostwald[39] "*I have been shivering since my youth, thinking of the possibility of disclosing the basis of integral science*".

Driven by his passion, Solvay embarked in the 1880s in the construction of an all-encompassing "*Gravito-Materialitic*" theory, based on an energetic concept of matter and gravity[40].

But what explains this devouring ambition in a man who didn't benefit from a real training in physics, and who didn't master its fundamentals?

Solvay was clearly one of those visionaries who grew up in the middle of the 19th century, and felt capable of solving the mysteries of the universe by mere reflection. Like Julius Mayer, a medical doctor and early formulator of the energy conservation principle[41], Solvay felt capable to discover the great laws of physics. One of his most striking features was his awareness of the value of his intuitions. Here was a man, who recognized a lack of detailed knowledge in physics and in chemistry, but who thought he could compete with professional scientists.

Solvay's astonishing self-confidence seems to go back to the beginning of the 1860's when he learned that the reaction he had in mind for the production of soda had been discovered fifty years earlier by Augustin Fresnel, the illustrious designer of the wave theory of light. This revelation may have forged Solvay's faith in the merits of his intuitions, giving rise to an assurance that would soon be reinforced by the success of his industrial fight. Thus, in the middle of the 1880's he had provided proof that he could make his ideas triumph... He had demonstrated his ability to surpass his predecessors[42].

Solvay's unwavering self-confidence, which astonishes us today, seems to find an echo in this reflection of Poincaré[43]:

> "*If I congratulate myself on industrial development, it is not only because it provides an easy argument for the advocates of science, but also above all because it gives the scientist the faith in himself.*"

Solvay's perseverance

Solvay's perseverance in his research bordered on obstinacy, a "*stubbornness*" as he called it, which pushed him to refute current physical interpretations and to seek the cause of all phenomena in the universal law of gravity[44].

Unfortunately, as we shall see, his incessant desire to reform physics in his own way, without taking advice from professional theorists, would lead to pointless research and to increasing scientific isolation.

But Solvay knew that he could count on the assistance of dedicated collaborators. In 1886, he called on the already mentioned Emile Tassel, a retired professor at Brussels' Free University, to complete his first essay[45]. Many years later, he extended this preliminary study and laid the foundations of his *Gravito-Materialitic* theory. He then called on younger collaborators. Among them, the physicist Edouard Herzen and the physical chemist Georges Hostelet[46]. We will have more to say on these two scientists in section 2.4.

In order to enlighten his new collaborators, Solvay provided them with typed notes[47], producing sketches of his ideas and records of his calculations. Here is part of a note "Science and the objective Universe", written on 15 September 1910:

> "*There are two clearly defined ways of practicing science: the exclusively experimental method and the exclusively interpretative method.*
>
> *But there is also a mixture of these two methods. The experimental method is above all suited for applications, and since man is incomparably more interested in practical uses than in pure concepts, and moreover since the progress in experimentation is incomparably easier to achieve than progress in philosophical conception, there was reason to predict that the latter would remain behind (...). For those who have greatly admired the design power of the ancient philosophers – especially the Greeks – who unlike us had almost no experimental facts to support their theories, it was clear that our time would witness the temporary bankruptcy of theoretical interpretation, immobilized or drowned by experimentation. I'm one of them, and I have predicted for about forty years the situation that we observe today. But at the same time, especially since 1880, and by a constant exclusively interpretative study of only a few dominant facts, I sought to fill the gap of bases which, in my opinion, would show up one day.*
>
> *I hasten to point out that it would have been difficult for me to practice science other than by remaining – and this by temperament – in the fundamental order of questions, since I do not know physics and chemistry in detail...*".

Despite his admitted shortcomings in physics and in chemistry, Solvay reassured himself, knowing that he could trust his intuitions. Some of his findings on heat-related issues were confirmed by professional scientists. Others became a source of inspiration for renowned specialists.

Here are some of Solvay's main results:
- 1858: Prediction of the existence of a matter-energy equivalence.
- Same year: Rediscovery of the law of Dulong and Petit on the thermal behaviour of solids (a law according to which all simple bodies have the same heat capacity).

- 1872: Discovery of the principle of the mechanical production of low temperatures; Same year: Realization of devices for the liquefaction of gases[48].
- 1886: Formulation of a theorem on the constancy of the cold produced in successive expansions (a result published in 1895).
- 1896: Rediscovery of a theorem of Mechanics (virial theorem).

Solvay's pioneering role

The originality of Solvay's early work was recognized by leading experts, who highlighted his pioneering role, notably in the production of cold[49]. We already mentioned the industrialist's idea that a mass could be attributed to energy. H. A. Lorentz had been struck by this remarkable prediction which dated back to 1858. Reacting to Solvay's death on May 26, 1922, he decided to write a note in tribute to the deceased.

This note[50], written with the help of Herzen, was presented to the Paris Academy on November 12, 1923. It contained the following comment:

"The views on which Solvay based his conclusions cannot be compared with a great fundamental principle like that of relativity; but it is nonetheless remarkable that a happy intuition allowed him to foresee, at a time when nothing required it, the existence of one of the most important relationships that modern physics has been able to establish."

Lorentz's opinion was shared by Augustin Boutaric[51], a professor at the Faculty of Science in Dijon, who celebrated Solvay's insight and his later attempts *"to confer on his opinion the certainty of a fact"*. These are his words:

"Many are the physicists who suspected the inertia of energy. Among the precursors one must mention Ernest Solvay, who in the 1880's undertook experiments in order to highlight it. He operated chemically on solid bodies endowed with strong reciprocal affinities (sodium oxide and phosphoric anhydride), and physically by producing repeated shocks inside agitated boxes. But he didn't note any appreciable change in mass. He simply concluded that the mass attached to the energies in play was below the experimental sensitivity...".

Impressed by the astonishing accuracy of the conclusion, Boutaric cited Solvay's reaction to the negative outcome of his experiments[52]:

"I didn't feel discouraged by my failure. Thinking about it, it became clear to me that the experiments carried out proved once again that the material equivalent of heat was not detectable under the conditions in which I had operated, nothing more. On the other hand, considerations of various kinds led me to the conclusion that the equivalent must be so small that reasoning alone, using calculation, should allow it to be determined, if however, it is determinable."

Today, we know that Solvay was right: it was by theory and not by experiment that Einstein realized in 1905 that mass and energy are linked by a famous equation, the validity of which would be definitively established in 1932 (thanks to the use of the Cockroft-Walton accelerator).

Let us emphasize that Solvay didn't just formulate theories. He always thought of the need to check their predictions and started looking for possible applications. This dual aspect of his activity is particularly striking[53]: Solvay tackled at the same time abstract questions of scientific philosophy and problems of a more practical nature.

Thus, driven by his interest in aviation, he decided to study the problem within the framework of his "*Gravitique des fluids*". In 1891, he concluded: "*The solution of the problem of aviation is near, this is my intimate conviction*".

In this respect, it is worth recalling the industrialist's lucid opinion about the future of airships. When asked in 1910 to write a preface to Robert Goldschmidt's book[54] on the subject, Solvay reacted as follows:

"I feel rather embarrassed. My friend M. Goldschmidt asks me to write a preface for his book "Les aeromobiles". But I find that having helped him to make the first Belgian airship is not a sufficient reason to justify my intervention. Because I don't believe in airships. On the other hand, I believe in aviation, and Mr. Goldschmidt knows it...".

We will come back to the author of the book, notably in section 2.2 and in the second part of our story. We will see that Solvay's reaction to Goldschmidt's request didn't cast any shadow on their friendly relationship.

Solvay's program and working method

The ways in which Solvay intended to react to the "bankruptcy of generally accepted theories" can be found in his (already mentioned) Note "*Science and the objective Universe*".

The industrialist's starting point was Kepler's third law, a rule he considered to be one of the most reliable in physics. Using planetary orbits, and enriching them with his "*universal Gravito-Materialitic law*" he set out to build an "objective" theory of the Universe.

But how did he proceed?

As a specialist in organization, Solvay chose a method suited to his peculiar situation. It involved four stages:

"First: formulate the fundamental principles. Second: draw all the consequences deductively. Third: submit the results to qualified collaborators, so that they can study them, complete them and correct them (if necessary). Fourth: tackle the question of experimental verifications."

Of these steps, only the first two were Solvay's own responsibility. For the others he relied on professionals: theoreticians and experimentalists. The former were his official collaborators: initially Tassel, later Herzen and Hostelet.

Experimentalists, on the other hand, had to be recruited on a case-by-case basis.

What can we say about Solvay's "four steps-method"?

Apart from the fact that the industrialist needed help for the two last steps, the method he adopted is intimately related to a well-known scheme: the "*Deductive method in Mechanics*". This is how Emile Picard, a French mathematician, described it in his book "*La science moderne et son état actuel*", published in 1905[55]:

"*Some illustrious physicists have chosen to break with old habits. Abandoning completely the historical point of view of the development of science, they adopt a point of view analogous to that of the geometer who constructs a geometry starting from a certain number of axioms. Their method is entirely deductive (...). An advantage is that the resulting system is well connected: one builds piece by piece an a priori set of interpretations, and one deduces from it all possible consequences. It is only after completion of the system that the results are compared with experiment...*".

In order to fulfil the last step of his program, Solvay needed an experimental facility. Taking example on Ludwig Mond[56], his industrial partner in Great Britain, he chose to set up a private laboratory at his home, *rue des Champs Élysées*. It is there that checks of his theories were carried out from 1892.

Unfortunately, the industrialist found himself confronted with a distressing fact: the severe lack of qualified physicists in his country... Hence, being unable to find local experts, he thought of a system of research subsidies[57] that would allow him to obtain the assistance of foreign experimenters.

The spring of 1910 marked the beginning of a new, promising phase in Solvay's scientific endeavours. Encouraged by his plan to boost the completion of his *Gravito-Materialitic* theory, he started revisiting his former Memoirs[58], written with Tassel's help, and called on the services of younger collaborators. Solvay now aimed at "*providing the science of the Universe with a simple representation by way of deduction, starting from firmly established postulates such as that which governs universal gravitation*[59]".

International recognition during his lifetime

Solvay's results in the field of low temperatures were recognized during his lifetime. This was notably the case with his discovery of a mechanical principle that enabled him to produce appreciable cold. Another achievement was his improvement of the Coleman process for cooling gases[30]. About ten papers, bearing his signature, appeared between 1895 and 1906 in the "Proceedings of the Paris Academy of Sciences".

In addition to his increasing reputation as a successful researcher, the industrialist-investigator benefitted from the impact of his donations: a scientific patronage that extended well beyond the borders of Belgium.

Solvay's generous gifts to faculties or laboratories in Paris, Nancy, Geneva, Berlin and Leiden, earned him respect and gratitude from all over Europe. They also led to notable distinctions:
- 1887: Solvay was elected corresponding member of the British Association for the Advancement of Science.
- 1899: He became an honorary member of the Royal Institution of Great Britain, and was invited to take part in the celebration of the Institution's centenary[61].
- 1907: Ostwald proposed his election as a member of the German Chemical Society.
- 1909: Solvay received the title of Doctor "honoris causa" in physical sciences from the University of Geneva[62].
- 1909: He was awarded the Leibniz Medal of the Prussian academy of Sciences, on proposal of Nernst and Emil Fischer.
- 1913: Solvay was elected corresponding member of the Prussian Academy of Sciences.
- 1917: He became a corresponding member of the Paris Academy of Sciences.

To conclude, we can say that Ernest Solvay was generally perceived as a scientific patron, who founded institutes and practiced his own research. Those who witnessed his devouring passion often regarded him as a stubborn autodidact, building theories and striving to confront them by experiment... But to these privileged witnesses, Solvay also appeared as an enlightened visionary who intuitively realized that physics was in crisis.

1.4 The quantum theory

Let us now look at this crisis, and let us see how it relates to the unexpected event of June 1911: the convening by Solvay of an "International Scientific Council".

Indeed, to explain the industrialist's action, it is necessary to examine the factors which caused it. This requires a brief reminder of an essential chapter in the evolution of physics during the years 1900–1910, known as the "first phase of the quantum revolution[63]". We will try to retrace its broad lines, limiting ourselves to the facts that have a direct link with our main subject.

The problem of black-body radiation

FIGURE 4: Max Planck in 1911, second rank (right hand side of the picture), see figure 1.

FIGURE 5: Wilhelm Wien in 1913 (centre of the picture), see figure 30.

Energy quanta are said to have been introduced in physics on December 14, 1900, with Planck's (figure 4) presentation of his radiation law[64] to the members of the German physical society. The introduction of these "*energy degrees*" was strikingly discrete. Planck had no idea that his announcement would cause a revolution.

The phenomenon of thermal radiation had been known for a long time. Physicists had observed that each body emits radiation, and that the colour of the light emitted by a white-heated body varied with temperature, gradually passing from red to yellow, then from yellow to blue... However, the characteristics of this transition didn't find a satisfactory explanation in the existing theories.

A formula, postulated by Wien (figure 5), seemed to agree with experiment in the region of *small* wavelengths. Unfortunately, it failed desperately at the other end of the spectrum (region of *large* wavelengths).

Making use of classical theories, Lord Rayleigh had produced a radiation formula which differed completely from that of Wien[65]. His result, known as the *Rayleigh-Jeans law*, seemed acceptable in the region of *large* wavelengths, but proved totally in fault outside this region... In short, the "black-body problem" seemed insoluble.

Planck's great achievement

Let us now take a closer look at Planck's announcement of December 14, 1900. The author of the new radiation formula – postulated at first, then proved – insisted on its capacity to represent the experimental data in all regions of the spectrum.

In other words, Planck knew that he had achieved an exploit.

But this awareness of success was not the main reason for his satisfaction. Something else seemed more important to him... Indeed, in his search for a justification of the formula, Planck had been able to rewrite the expression in terms of two *universal constants*, one of which came from the theory of gases[66] – he designated it with the letter k. This meant that the new law could be used to obtain the value of k, starting from radiation-data, *i.e.*, data which related to light waves propagating in the *ether*. Thus, knowing the value of k, Planck could make use of well-known relations and obtain the value of fundamental *molecular* quantities, such as Avogadro's number and the electrical charge of the electron. In other words, he had succeeded in establishing a new link between two distinct physical worlds: the world of *matter* and the world of *radiation*.

It is this "tour de force" that Planck took care to emphasize when he spoke at the German physical society. On the other hand, he didn't elaborate on the price he had been forced to pay for this feat: the need to base his demonstration on an unheard-of hypothesis, in flagrant contradiction with the laws of mechanics: the quantization of the energy of a set of material oscillators[67].

Planck's hypothesis of "indivisible units of energy" – they would later be called quanta – led him to postulate the existence of a *universal constant*, alien to all known theories: a constant of proportionality between the energy of an oscillator and its oscillation frequency. Planck designated it by the letter h.

But what was the meaning of this new constant? Planck was unable to tell...

However, it seemed clear to him that h could only belong to the microscopic world of matter, a mysterious territory in which a key-element, the electron, had only recently been identified.

There was one point that Planck took for granted: the fact that the hypothesis didn't affect the propagation of light. According to him, the constant "h" intervened only in the absorption and emission of radiation by matter. Determined to save "what might be saved", Planck refused to question the classical Maxwell–Lorentz theory of light, and the continuous, undulating nature of light, so vividly confirmed by experiment.

Reactions to the quantum hypothesis

Curiously, Planck's announcement of December 1900 didn't arouse great interest. Nobody even thought of attacking his theory. Other, more burning questions were on the physicists' minds: the discovery of the electron, the mysterious X-rays and the still puzzling radiations emitted by radioactive bodies.

However, several questions about Planck's theory would be addressed during the first decade of the twentieth century:
- Was the quantum hypothesis a necessary condition, or just a convenient assumption capable of justifying Planck's radiation law?
- Could the presence of quanta be detected in other measurable phenomena than those related to thermal radiation?

FIGURE 6: Hendrik Antoon Lorentz in 1915, Academisch Historisch Museum Leiden.

A first reaction to Planck's work appeared in 1903. It came from Lorentz (figure 6), whom we already met. This leading theoretician, famous for his *electron theory*[68], wrote a paper on the emission and the absorption by metals of large wavelength radiation. According to the electron theory, this radiation was due to the motion of free electrons, circulating in the metal. Having obtained a distribution law[69] for the corresponding radiant energy, Lorentz noticed that the energy distribution for large wavelengths was the same as that given by Planck, a fact all the more surprising as the theoretical deductions were totally different. He mentioned Planck's hypothesis about the oscillators' energy, which had to be a multiple of a finite quantity, and observed that it should be regarded as an essential element of the theory. However, Lorentz didn't seek to grasp its significance.

The role of Planck's quantum of action – *i.e.*, his constant h – as well as other essential elements of his theory became the subject of detailed studies, notably by Einstein[70], Ehrenfest[71] and Lorentz.

Lorentz's study culminated in a demonstration that Planck's law had no place in the edifice of classical theories. He presented his conclusion in Rome at the Congress of mathematicians in 1908 (a presentation which, as we will see, would have unforeseen consequences for Planck). Lorentz's analysis, written in French[72], appeared in "*La revue générale des sciences pures et appliquées*". It aroused the admiration of Einstein who expressed his feelings in a letter to Lorentz in 13 April 1909. This letter[73], together with a previous one[74], to which Lorentz replied[75] at length, marked the beginning of an ongoing correspondence between the two men, and of a deep friendship, which only ended in 1928 with Lorentz's death.

Other consequences of Planck's theory were searches for phenomena suggestive of a quantization of energy. These included the photoelectric effect and the puzzling behaviour of specific heats at low temperatures. We shall be specially concerned by the specific heat problem, because of its direct link with the 1911 Solvay Council.

But let's proceed step by step, and focus on the events that took place during the few years that followed Lorentz's reaction in 1903.

Einstein's first contribution

A crucial event took place in March 1905: the publication by *Annalen der Physik* of a paper signed by Albert Einstein (figure 7), an official attached to the Bern patent office.

In this paper[76] entitled "*On a heuristic point of view concerning the production and transformation of light*", the author developed a point of view, based on a bold hypothesis: the existence of *quanta of light*.

Einstein's revolutionary idea owed nothing to Planck's theory. It resulted from the analysis of an experimental fact: the behaviour of thermal radiation in the region of small wavelengths.

Einstein was struck by the contrast between the customary description of matter, in terms of *discrete* atoms, and that of light, in terms of *continuous* electromagnetic waves. Deeply troubled by this lack of unity, he re-examined the classical concepts and came to the conclusion that their application to black-body radiation resulted in an absurdity: the presence of an infinite quantity of radiant energy in a limited portion of space.

FIGURE 7: Albert Einstein in 1911, see figure 1.

Hence, being unable to rely on the classical theories, Einstein turned to experiment, and focused on the small wavelength radiations (*i.e.*, radiations belonging to the high frequency region, source of the above absurdity) which happened to obey Wien's law. He therefore decided to submit this law to a thorough examination and came to the conclusion that radiations of a given frequency, taken within the region of validity of Wien's law, behaved as a "gas" of quanta, the energy of which was proportional to the frequency.

The result of the study was clear, but total mystery remained. How could the presence of "energy quanta" be reconciled with the *continuous* wave nature of light, a fact so powerfully confirmed by experiment?

The light quantum hypothesis was all the more disturbing because it appeared as a return to the since long rejected *corpuscular* concept of light... Einstein therefore decided to test the light quantum hypothesis by applying it to *high frequency* phenomena that couldn't find a simple explanation in the customary theoretical framework. He chose the photoelectric effect[77], a phenomenon in which electrons are ejected from a metal under the impact of *ultraviolet* light, and showed that the hypothesis could lead to an explanation of its characteristics. Einstein obtained in this way an experimentally verifiable "photoelectric equation". Unfortunately, its verification would take about ten years (Robert Millikan's results of 1914).

Reactions to Einstein's light quanta

Though quite promising, Einstein's result did not bring the expected success. The most authorized experts, with Planck in the lead, remained firmly opposed to the idea of light quanta.

Planck would recall his feeling about the matter in a letter of 13 June 1913, which aimed at Einstein's election as a full member of the Prussian Academy of Sciences. These were his words[78]:

> "That Einstein sometimes got lost in speculations which failed their goal, as in the case of the light quanta hypothesis, should not be blamed on him with too much rigor...".

The opposition to light quanta would last for more than fifteen years. It was only in 1923, nine years after Millikan's decisive results[79] that the validity of the photon-concept (a concept which resulted from extending the idea of an energy quantum to that of a *particle*) would be confirmed, thanks to the discovery of the Compton effect, and to its interpretation.

Going back to the photoelectric equation of 1905, there is a point which is worth mentioning. Documents show that, after reading Einstein's revolutionary paper, Lorentz thought of starting an experimental study of the photoelectric effect[80]. This fact, which remained largely unnoticed, is particularly revealing. It tells us that the author of the electron theory was one of the few physicists who kept an open mind and who distanced himself from Planck's conservative position[81]. However, Lorentz's project, conscientiously matured, encountered an obstacle that we will have ample opportunity to evoke... Being overloaded with responsibilities he hadn't foreseen, the great theorist found himself unable to devote full attention to the photoelectric effect. He gave up his plan when he learned of Millikan's first results.

The specific heat challenges

Let us now focus on the event which triggered Einstein's second contribution in relation to quanta: Planck's publication in 1906 of his *"Lessons on heat radiation"*.

After reading this work, the author of the light quantum hypothesis came to the conclusion that there was no disagreement between Planck's concepts and his own, but that Planck's reasoning suffered from an internal contradiction that could only be lifted at the cost of an additional assumption, equivalent to the idea of light quanta.

This point being established, Einstein set out to test the reality of Planck's quantized oscillators. He reasoned as follows[82]:

> "*If Planck's theory of radiation strikes to the heart of the matter, then we must also expect to find contradictions between the present kinetic-molecular theory and experiment in other areas of the theory of heat, contradictions that can be resolved by the route just traced...*".

Einstein knew of the contradictions that existed between the thermal properties of solids and the predictions of kinetic theory[83]. It had been shown that at low temperatures the specific heats[84] of various solids didn't obey the law of Dulong and Petit. It therefore seemed imperative to examine whether agreement between theory and experiment could be re-established by treating each atom of the solid as a "Planck-oscillator". Einstein chose to work within the framework of a simplified model. Assuming that all atoms of the solid oscillate around their equilibrium position with the *same* frequency, he obtained a specific heat formula which agreed with the classical expression at high temperatures, but predicted a decrease of the specific heats at low temperatures and their vanishing at the approach of the absolute zero. This meant that the new formula could possibly explain the non-classical behaviour of specific heats at low temperatures.

It was a first and encouraging step. But Einstein thought further. He realized that the predictions of his "quantum formula" could be verified, in principle, by measuring specific heats at sufficiently low temperatures. If conclusive, these measurements could provide decisive support to Planck's hypothesis, by establishing its validity in a domain that was *not* the domain of black-body radiation.

However, Einstein's enthusiasm was tempered by his awareness that at very low temperatures the specific heat values were very small, which made their measurement extremely delicate. In other words, the validity of Planck's quantum hypothesis seemed still far from being established...

What Einstein didn't know was that a team of chemists was about to embark on a campaign of such measurements, for reasons quite different from his own.

An unexpected actor

The team in question was led by Walther Nernst (figure 8), the director of the Institute of Physical Chemistry at the University of Berlin.

Nernst was above all the author of an ambitious, but still highly controversial "*Heat Theorem*". This proposition of 1906 should allow the calculation of the equilibrium conditions of a chemical reaction by means of thermal data[85]. If confirmed, it would be a definite step towards a Nobel Prize... Excited by the prospect, Nernst decided to do everything to validate his "theorem".

FIGURE 8: Walther Nernst in 1911 (sitting, left hand side of the picture), see figure 1.

The proposal predicted a decrease of specific heats with temperature, and their convergence towards a same limit in the neighbourhood of the absolute zero. To check if this was the case, Nernst engaged his team in a "low temperature measurement campaign". At the same time, he started looking for theoretical elements that might support his claim.

Was there any recent theory of specific heats, capable of explaining the observed deviations from the law of Dulong and Petit?

It was only at the end of 1908 that Nernst got informed of Einstein's work. Two facts seem to indicate that it was by the reading of an article by M. Thiesen[86].

On the one hand, there is a paper by Nernst[87], presented at the Berlin Academy on 21 January 1909, which contains a reference to Thiesen's article, where it is said that Einstein "*could be the author of a promising theory of specific heats*". The article describes the main lines of the theory. Thiesen mentions Planck's radiation formula, and the use that Einstein may have made of it... But he says nothing about quanta.

On the other hand, we notice that two papers – one by Einstein, the other by Thiesen, are cited jointly in the sixth edition[88] of Nernst's *Theoretische Chemie*, which appeared in 1909. The edition contains two sentences that are identical with Thiesen's, but it doesn't mention Planck's name...

We will come back to the puzzling absence in Nernst's writings of any reference to the quantum theory (see section 1.6).

However, one thing is sure: as soon as Nernst got informed of Einstein's work, he took a close look at it, and realized that the new formula supported his theorem.

He immediately decided to check the formula's validity, by measuring specific heats at lower and lower temperatures, down to the boiling point of air. The stain was heavy. Nernst had to wait until the end of 1909 to collect the first results. But his patience was rewarded: the measured values appeared to be in line[89] with Einstein's predictions.

1.5 How Nernst discovered Einstein's genius

Enchanted by the success, Nernst hastened to Zurich to meet the author of the providential formula. The encounter took place in the beginning of March 1910.

Einstein was, at the time, associated professor at the cantonal University of Zurich, his earliest full-time academic position.

The visit of *Geheimrat* Nernst, a leading authority of the University of Berlin, made a big impression on Einstein's colleagues[90]. The young associated professor informed his visitor about the origin of his formula: Planck's quantum theory.

He also told him that a verification of its predictions at sufficiently low temperatures would provide proof of the validity of the quantum hypothesis, urging him to continue the effort by pushing his measurements down to the boiling point of hydrogen.

For Nernst, the interview with his young colleague was a real revelation. Deeply struck by Einstein's personality, he expressed his feelings in a letter to Arthur Schuster, declaring[91]:

"At the moment I am in Lausanne, partly in order to get a breath of fresh air, and partly to brush up my French, but tomorrow I am already returning home.

On my trip here I visited Prof. Einstein in Zurich. It was for me an extremely stimulating and interesting meeting. I believe that, as regards the development of physics, we can be very happy to have such an original young thinker, a Boltzmann "redivivus"; the same certainty and speed of thought; great boldness in theory, which however cannot harm, since the most intimate contact with experiment is preserved. Einstein's "quantum hypothesis" is probably among the most remarkable thought constructions ever; if it is correct, then it indicates completely new paths both for the so-called "physics of the ether" and for all molecular theories; is it false, well, then it will remain for all times a beautiful memory...".

Einstein was also very happy with Nernst's visit. He told his friend, Jacob Laub[92]:

"For me, the theory of quanta is a settled matter. My predictions regarding the specific heats are apparently being brilliantly confirmed. Nernst, who has just been here to see me, and Rubens are busily engaged in the experimental verification, so that we will soon know where we stand. As for the quanta, I have already discovered something interesting but nothing final...".

Einstein's last remark was apparently an allusion to the mystery of light quanta, a persistent enigma about which he had expressed his dismay in a previous letter[93]:

"...This quantum question is so extraordinarily important and difficult that everybody should take the trouble to work on it...".

According to Einstein, the only encouraging point about quantum theory was the fact that the predictions of his specific heat formula seemed to agree with low temperature data, a promising result which would hopefully be confirmed. As to the more serious problem of radiation, Einstein had recently obtained indications[94], suggestive of a "wave-particle" nature of light... But these results, in striking contradiction with Maxwell's predictions, were more intriguing than ever. Deeply disturbed by the situation, Einstein embarked on the construction of a new electromagnetic theory which would account for the existence of electrons and for the "dual" nature of light. This attempt absorbed all his energy between March 1909 and May 1911, but proved unsuccessful. Here are messages to three colleagues which attest of Einstein's failure and of his discouragement:

- Einstein to Johannes Stark[95] in the summer of 1909: *"You cannot imagine how hard I have tried to contrive a satisfactory mathematical formulation of the quantum theory. But I have not succeeded thus far...".*
- Einstein to Sommerfeld[96] in July 1910: *"I hesitated so long to write to you because, on the one hand, I would so much like you to come to Zurich during the summer vacation, but, on the other hand, I lack the courage to urge you to do so. For I have not been able to come up with anything even partly complete regarding the constitution of radiant energy...".*
- Einstein to Laub[97] in August 1910: *"... I have not made any progress regarding the question of the constitution of light. There is something very fundamental at the bottom of it...".*

As we already mentioned, the reality of light quanta, and of *photons*, would be confirmed in 1923. Yet, the true nature of a *photon* would continue to haunt Einstein until the end of his life. These are his words in a letter of 1951 to his friend Michele Besso[98]: *"Fifty years of meditation didn't provide an answer to this simple question: what is a quantum of light?".*

1.6 Nernst's great dilemma

Let us now return to the beginning of March 1910, and let us re-examine Nernst's letter to Schuster. The author's surprising freedom of tone and his enthusiasm for Einstein's quantum hypothesis, are in sharp contrast with his concern to avoid any allusion to quanta in his public statements...

Nernst's discretion with regard to Planck's theory persisted until January 1911. It was obvious in his lecture of April 1 of 1910 at the French Physical Society. Nernst spoke of his heat theorem, announced the results of his measurements, and went so far as to mention Einstein's theory[99]... But he didn't say a word on Planck, nor on quanta.

Nernst's secretive attitude was still notable in May 1910 during his visit of the Bunsen Society in Giessen[100]. He evoked his theorem, cited the specific heat formula, specifying that it had been established by Einstein *on the basis of Planck's law*, and highlighted the agreement between the results of his measurements and the new formula (after a minor empirical correction). This time, Planck's radiation law was mentioned, but his theory stayed in the shadow... Still nothing about quanta.

Why this strange restraint?

What Nernst didn't tell Schuster was his perplexity following Einstein's revelations.

On the one hand, he felt reassured, imagining that the heat theorem could be derived from Einstein's specific heat theory[101]... On the other hand, he felt uneasy about the existence of a close link between his theorem and Planck's theory, a theory he used to describe as an *"incongruous"* proposition, calling it at best "*a simple interpolation formula*[102]".

For Nernst, this "quantum connection" was a real source of risk. Everyone in Berlin knew that Planck had been denied a Nobel Prize in 1908 on the pretext[103] that his theory was based on a *"completely new and barely plausible hypothesis, namely the existence of an elementary quantum of energy"*.

The news had been a snub for Planck, who had been told that he would get the Prize, and had accepted to grant interviews to journalists, who reported his reaction in the newspapers[104].

The incident caused a stir at the time. It had been orchestrated by the Swedish mathematician Gösta Mittag-Leffler, who had heard that the Committee's decision in favour of Planck had been advocated by Arrhenius, his colleague and powerful rival. Firmly decided to oppose the proposal, Mittag-Leffler, who didn't belong to a Nobel Committee, chose to act as a member of the Stockholm Academy. Knowing that the Committee's proposal had to be approved by the Academy, a procedure which usually amounted to a simple formality, he decided to reveal Lorentz's reservations about the validity of Planck's law, a scepticism he had witnessed a few months earlier in Rome as he attended the Mathematician's Congress.

We already mentioned Lorentz's statement at the Congress that Planck's law had no place in the edifice of classical theories. Indeed, in his report *"On the Distribution of Energy between Ponderable Matter and the Aether"*, Lorentz had shown that the classical equipartition of energy led inexorably to the Rayleigh-Jeans law, and that Planck's law could only be accepted at the cost of a radical reform of the usual concepts. His conclusion that new experiments were needed to decide between the two laws made a strong impression. It also aroused strong reactions, notably from Wien[105] and from other physicists, who had no doubt about the validity of Planck's law.

Whatever the relevance of Lorentz's reservations, they seemed sufficient to Mittag-Leffler, who mentioned them during the Academy's full session, thus blocking the Nobel Committee's proposal[106].

Ironically, on 10 December 1908, date of the Nobel ceremony, Planck was neither in Stockholm, nor in Berlin. Invited to deliver a talk at the University of Leiden, he was staying with Lorentz.

Chapter 2
An unprecedented project

Let us now turn our attention to the next events that happened during the spring of 1910. We easily imagine Nernst's perplexity after his encounter with Einstein. He suddenly realized that the acceptance of his theorem was intimately linked to the fate of an "uncertain and contested theory". What could he do to break the deadlock?

2.1 The idea of a Council

It seems that an idea came to Nernst's mind in April 1910, shortly after his Parisian presentation. A possible solution was to facilitate a validation of quantum theory by reorienting the debate towards the field in which Planck's hypothesis had recorded an indisputable success: the molecular domain.

The idea of drawing attention to Einstein's specific heat theory seemed all the more judicious, since Frederick Lindemann, Nernst's British assistant, had succeeded in establishing a link[107] between the only parameter in Einstein's formula (a vibration frequency[108], characteristic of the solid) and other measurable quantities, such as the solid's density, its melting point, and its atomic weight...

Nernst had no doubts. It was time to bring the failure of classical molecular theory to light, and to publicize the success of Einstein's quantum alternative. This required the convening of an international "Scientific Council[109]".

But where should the Council take place?

The natural location was Berlin, cradle of quantum theory and seat of Nernst's Institute of Physical Chemistry. However, it was clear that the German capital wasn't suited for an international "summit" which aimed at confirming the supremacy of a German theory... Hence, it was imperative that the Council took place on a neutral ground. Moreover, it seemed preferable that it be convened by a non-German personality, alien to Nernst's professional sphere.

Indeed, the Berlin chemist couldn't run the risk of being seen as a scientist in search of a Nobel Prize. More than any candidate, he had to

FIGURE 9: Svante Arrhenius in 1922, First Solvay Council on Chemistry, see figure 60.

keep a low profile due to his long-standing conflict with Svante Arrhenius (figure 9), an influential member of the Nobel Committee for physics, and one of the few Swedish scientists whose opinion was authoritative[110] in the Committee for chemistry.

The Nernst–Arrhenius conflict

Arrhenius and Nernst had been close during their student years. They had met at Friedrich Kohlrausch's Institute in Würzburg, and had spent time with Ludwig Boltzmann in Graz.

Unfortunately, the early comradeship had since long given way to extreme rivalry. Both men were ambitious. Nernst used to say that Sweden was not conducive to the development of a scientific career, a fact of which Arrhenius had provided the best proof...

Arrhenius, for his part, blamed Nernst for having enriched himself by the sale to the AEG Company of his electrolytic bulb. He said that by doing so, Nernst had offered the world *"a rare example of a commercial transaction between a researcher and a company, the effects of which had not been beneficial to the Company"*. In addition, he accused his German colleague of having enhanced the proceeds of the sale by investing it in nightclubs[111].

An attempt to reconciliation had taken place in 1897, on the initiative of a mutual friend. A meeting of the two men had been planned at Stockholm's Grand Hotel. Nernst, who had come with his electric bulb, couldn't resist the temptation to show its performance. He plugged it in, and blew all the fuses in the hotel... Arrhenius, had the misfortune of not being able to contain himself: he had a fit of giggles that signalled the end of the attempt to reconciliation.

Another major blow for Nernst was the announcement in 1903 that Arrhenius would receive the Nobel Prize for chemistry. His resentment redoubled when he learned that there was talk of creating an Institute of Physical Chemistry in Stockholm, and of appointing Arrhenius as its leader. Yielding to anger, Nernst accused Arrhenius of having embezzled funds from the Nobel Foundation *"to become the director of a Swedish Institute, erected with stolen money"*.

Arrhenius didn't lay down on it. He used his influence on the members of the two Nobel Committees to block Nernst's claims at a Prize, explaining that Germany had enough nightclubs[112] and that the Nobel Foundation shouldn't be used to create more...

An unprecedented project 39

We know that Arrhenius had the last word. Responsible for writing the final report on Nernst's nominations, he managed to freeze his award for nearly fifteen years, not hesitating to say:

"*Even if the heat theorem were found to be confirmed there would be moral reasons not to award a Nobel Prize to its author*[113]".

2.2 A providential man: Ernest Solvay

Aware of the difficulties that awaited him in Stockholm, Nernst was determined to take up the challenge. The Council, the plans of which he was forging, would be his best weapon against Arrhenius.

But the author of the heat theorem still had to find the scientific patron who would be willing to convene the Council. He suddenly remembered Ernest Solvay, the Belgian industrialist who had come to Berlin a few months earlier to receive the Leibniz Medal of the Prussian Academy of Sciences (a distinction granted to him on Nernst's and Fischer's proposal). Here was a man who met all the criteria, and with whom Nernst had celebrated on July 1, 1909, the presentation of the prestigious Medal at Berlin's Automobile Club:

— Solvay lived in Brussels, world capital of internationalism (out of 112 permanent international associations in 1906, 17 had no fixed headquarters, 15 were located in Switzerland, 2 in the Netherlands and 42 in Belgium[114]). As witnessed by Ostwald, *internationalism was seen and nurtured in Belgium as an indigenous industry*[115].

— A world exhibition was taking place in Brussels and a large number of international Congresses would be held in the Belgian capital from 23 April to 1 November 1910... Among them, an "International Congress of Radiology and Electricity", planned for September 1910, which would be attended by illustrious radioactivity experts, such as Marie Curie and Ernest Rutherford. In short, Nernst's Scientific Council would be in line with a series of prestigious meetings[116].

In addition to these considerations which favoured Brussels, there were important factors which pleaded for Solvay. Nernst had been in contact with the industrialist for more than ten years (their path had crossed in London in 1899 on the occasion of the centenary of the Royal Institution of Great Britain). He remembered Solvay's speech at the Automobile Club, his passion for physics, his interest in the constitution of matter and energy, and his moving confession of being "*haunted by the Truth to be discovered*".

There could be no doubt: Solvay was much more than a founder of Institutes. He would, in all probability, make a point of lending his assistance to a German friend who planned an operation which aimed at resolving a major crisis in physics.

A final, decisive advantage in choosing Brussels was that Nernst could count on the support of his Belgian assistant, Robert Goldschmidt, a young chemist, closely linked to the Belgian financial world[117], whom he had recently promoted to deputy director of his laboratory for experimental thermodynamics[118].

Above all, Goldschmidt had a home in Brussels... and happened to be a friend of Solvay. He was obviously the man who could take care of the organization of the Council.

Robert Goldschmidt, inventor and chemist

Born to parents who had come from Germany, Goldschmidt (figure 10) became a Belgian citizen in 1896. At the end of his studies at Brussels' Free University, he obtained the title of doctor in science, chemistry section. After a stay with Nernst in Göttingen, he returned to Brussels and defended a thesis *"On the relationship between dissociation and thermal conduction in gases"*.

FIGURE 10:
Robert Goldschmidt in 1911 (standing, left hand side of the picture), see figure 1.

One of Goldschmidt's ambitions was to popularize science by familiarizing the public with electricity. Driven by his taste for invention, he ventured into a variety of fields, including the design of wood-burning trucks for the Congo and the development of wireless telegraphy[119]. In 1906, he joined Paul Otlet, a bibliography enthusiast about whom we will have more to say in section 2.6, and developed with this pioneer in information science the "Bibliophote[120]", a prefiguration of the microfilm.

Goldschmidt then turned his attention to a very different problem. With Solvay's help, he ventured into building Belgium's first airship[121].

Oddly enough, the man didn't feel satisfied with his role of inventor. Seeking to climb the steps of an academic career, he approached Nernst, his former mentor, and became his deputy in a Berlin laboratory[122]. It is in this capacity that Goldschmidt took part in Nernst's campaign of specific heat measurements. His merits were recognized by the author of the heat theorem, who, commenting on his results in Paris on April 1, 1910, declared[123]:

"These measurements have been made in the thermodynamics section of my laboratory, headed by Dr. Robert Goldschmidt and by myself."

An unprecedented project   41

2.3 Nernst in action

Let us come back to the Council. In order to prepare the operation, Nernst revealed his plan to Goldschmidt. Following Einstein's advice, he took care to strengthen his position by pushing his measurements towards lower and lower temperatures. In order to do so he went to Leiden to see Kamerlingh Onnes, alias "Mr. Absolute Zero", the man who succeeded in liquefying helium in 1908.

In June 1910, Nernst felt ready to launch the operation "Council". He drew up a list of members to be invited[124], and submitted his plan to three colleagues: Planck, Lorentz, and the Danish physicist Martin Knudsen.

Planck's reaction

Planck expressed reservations about a meeting he considered premature. This is his letter to Nernst[125] of June 11, 1910:

> *Dear Colleague,*
>
> *Please allow me to add some general remarks to the marginal notes that I made on your manuscript with your consent.*
>
> *Your idea corresponds entirely, in magnitude, to the problem which it aims to solve and, in this respect, I can only agree with it with the deepest conviction.*
>
> *But I cannot hide my great fear with regard to its feasibility. I already indicated in my marginal notes that the result of a Congress of this kind would be more considerable in my opinion if it could be postponed a little longer, so that we could have more extensive data.*
>
> *However, there is in my opinion another point which should prompt us to wait another year. Indeed, the convening of the Congress is based on the presumption, as your statement of reasons highlights appropriately, that the state of the theory, generated by the laws of radiation, by the specific heats, etc. is full of gaps that are unbearable for any theorist, therefore requiring a joint action to find remedies.*
>
> *On the other hand, relying on my experience, I am convinced that barely half of the participants you have in mind realize the absolute necessity of a reform, with a sufficient acuteness to feel inclined to attend the Congress. As for the older ones (Rayleigh, van der Waals, Schuster, Seeliger) I don't know if they would even get mildly enthusiastic about the project. Among young people, too, the urgency and importance of these issues are far from being sufficiently recognized. Among those you mention, it seems to me that only Einstein, Lorentz, W. Wien and Larmor are seriously interested in the matter... But let a year pass, or better two years, and you will see how*

> much the gap, that only begins to open up in the theory, will open more and more, so that those, who are still far away today, will finally be dragged into the problem.
>
> Needless to add, I presume, that whatever is done in this direction, I will take the keenest interest in it, and that I promise in advance my full collaboration in any undertaking of this kind. Indeed, I can say without exaggeration that for ten years nothing has excited me more in physics, and attracted me more powerfully than these quanta of action...

Reassured by Planck's final words, Nernst chose to ignore his reservations and decided to go ahead. His main concern was to obtain Solvay's agreement...

What would be the industrialist's reaction to the idea of a *Scientific Council*? Would he accept to sign the invitation letters? The only way to find out was to go to Brussels and to meet Solvay. Nernst instructed Goldschmidt to organize an encounter at his home.

A decisive meeting

The encounter took place at the end of June, or in the beginning of July 1910[126]. We have no record of what was said at Goldschmidt's. However, we may assume that Nernst avoided revealing his plan, and that he contented himself with probing Solvay. This seems clear from Nernst's personal message to Solvay[127] of July 26, 1910:

> Very honoured Mr. Solvay,
>
> I hereby allow myself to submit a proposal to you, and at the same time I would ask you kindly – in case, for any reasons, you do not wish to take care of the matter – to instruct the wastepaper basket to respond to the letter and to its appendix.
>
> But you might be interested in the project, and I speak to you, my very honoured Mr. Solvay, because I know how much you care about general and important problems, and also because for an "international Council", as I would like to suggest to you, Brussels seems more appropriate than Berlin, Paris or London. The particulars can be found in the included invitation draft, and my friend Goldschmidt, with whom I have discussed the project in depth, will be more than happy to give you more detailed information.
>
> The Council will only work usefully if the participants are exclusively researchers who are specially concerned by these problems, and who have a keen interest in them.
>
> I allow myself to suggest the following invitees:

Honorary chairman: Solvay
Chairman: Lord Rayleigh (England)
Secretaries: Dr. Goldschmidt and a younger person.
Members: Einstein (Switzerland), Knudsen (Denmark), Hasenöhrl (Austria), Lorentz (Holland), Langevin, Perrin (France), van der Waals (Holland), Larmor, Jeans, Schuster, J. J. Thomson, Rutherford (England).

So far, I have only talked about the project confidentially to Lorentz, Knudsen and Planck, without specifying neither time nor place. All these gentlemen would gladly collaborate[128], so that, judging by these first contacts, the success of the Council is hardly in doubt.

Nernst also sent a second, *confidential* letter. It was the draft of a letter of invitation that Solvay would be asked to sign in the event of an agreement[129]:

INVITATION TO AN INTERNATIONAL SCIENTIFIC COUNCIL TO ELUCIDATE SOME CURRENT QUESTIONS REGARDING THE KINETIC THEORY

By all appearances we are at the moment in the middle of a new evolution of the principles on which the classical molecular kinetic theory of matter is based.

On the one hand, this theory, in its well-reasoned development, leads to a radiation formula that disagrees with all experimental results. On the other hand, from the same theory arise predictions on the behaviour of specific heats (law governing the variation of the specific heat of polyatomic gases with temperature; validity of the rule of Dulong and Petit at low temperatures) which are also refuted by numerous measurements.

As has been shown in particular by Messrs. Planck and Einstein, these contradictions disappear when certain limits are imposed on the motion of electrons and atoms in the event of oscillations around a rest position (assumption of energy degrees). However, this idea, in turn, deviates so considerably from the equations of motion of material points that have been used so far, that its acceptance would necessarily and unquestionably require a major reform of our fundamental theories.

The undersigned, although alien as a result of his other work to special questions of this kind, but animated with a sincere enthusiasm for all problems which by their study broaden and develop our knowledge of nature, believes that a written and verbal exchange between researchers more or less directly concerned by these problems, could, if not lead to a final decision, at least clear the way to their solution.

A big step on the way to the continuous and smooth development of atomistic theory would already be taken if one could clearly establish which of our molecular and kinetic interpretations agree with the observation results, and which, on the contrary, should undergo a complete transformation.

> *To this end, the undersigned invites you, among others, to participate in a "Scientific Council" to be held in Brussels around ......, bringing together a select few eminent professionals.*
> *After kindly indicating their acceptance, the members of the Council will receive the following conference topics, each of which will undoubtedly be the subject of an in-depth discussion:*
> 1. *Deduction of Rayleigh's radiation formula.*
> 2. *To what extent does the kinetic theory of ideal gases agree with experimental data?*
> 3. *The kinetic theory of specific heats according to Clausius, Maxwell and Boltzmann.*
> 4. *Planck's radiation formula.*
> 5. *The theory of energy quantities (degrees).*
> 6. *Specific heats and the "theory of degrees".*
> 7. *Consequences of the "theory of degrees" for a series of physical and physicochemical problems.*
>
> *It is intended to publish the reports at a later date, together with extracts of the discussions, all in one volume.*
>
> *Signed E. Solvay*

The above letters are interesting for a number of reasons. In the first place, they show that Nernst proposed a scientific meeting of a totally new type. The Council he had in mind was very far from a usual Congress. Nothing to do with the one held in Paris in 1900, which had been convened with the purpose of drawing up a report on the general state of physics. Nernst's idea was to bring together a small group of experts, from different countries, in order to take stock of a well-defined question. This method, after having proved successful in 1911, would become the trade mark of the famous Solvay Councils. Recognized as an absolute reference, the method would be adopted in Europe and in America.

The first conference organized on the same model was the one that took place in 1913 in Göttingen[130]. After the Second World War (1947), the method of the Solvay Councils would be adopted in the United States for the first post-war international conferences in physics[131].

Another noteworthy feature of the second letter is the title suggested by Nernst: "*International scientific Council to elucidate some current questions regarding the kinetic theory*". Note the difference with the title of the Council's official report[132], published in 1912: "La théorie du rayonnement et les quanta".

Nernst clearly chose a title which suited his priorities; hence the emphasis on the problems of classical molecular kinetic theory.

Furthermore, we may assume that Nernst's choice was partially motivated by his wish to arouse the industrialist's interest. There is little doubt that the failure of molecular kinetic theory was raised during the conversations at Goldschmidt's, and that Solvay informed Nernst of his rejection of a theory he regarded as a mere *fantasy* (see section 2.8).

Last but not least, we notice the absence in Nernst's invitation letter of a list of the scientists who would be invited[133] (an anomaly to which we will come back in section 2.5).

List of invitees, designation of a chairman

Contrary to the Council's agenda, which may have been established in consultation with the three scholars cited in the first letter, it is clear from Planck's reaction that the list of Council members had not been subject to prior agreement.

In fact, a simple look at Nernst's proposal tells us that several members were chosen for their notoriety. We notice, in particular, the presence of five holders of a Nobel Prize: Röntgen, Lorentz, Rayleigh, J. J. Thomson and Rutherford.

In other words, Nernst's desire to surround himself with influential physicists had led him to ignore the condition formulated in his letter to Solvay:

"*The Council will only work usefully[134], if the participants are exclusively researchers who are specially concerned by these problems and who have a keen interest in them.*"

It was clear to any professional physicist that neither Röntgen, nor Seeliger were meeting this criterion.

Conversely, Nernst's idea of entrusting the Council's presidency to Lord Rayleigh was natural and amply justified:
— Rayleigh, a physics "giant", belonged to the British scientific elite.
— He had chaired the Royal Society, and had directed Cambridge's Cavendish laboratory until 1884.
— He was known for having made an in-depth study of black-body radiation.
— He had recently been elected corresponding member of the Berlin Academy, a nomination which had brought him in closer contact with German scientific circles.

Furthermore, Nernst may have been motivated in his choice by more personal reasons... Indeed, Rayleigh was the discoverer with William Ramsay of argon, a noble gas, the study of which could provide evidence[135] in support of the heat theorem.

It was also Rayleigh, who delivered in 1899 the presidential talk at the Royal Institution on the occasion of its centenary, a celebration in which Solvay had taken part.

Solvay's reaction to the proposal

Solvay answered on 5 August 1910. He told Nernst that he agreed in principle with his proposal, but that he wished to postpone the Council until the fall of 1911 (Nernst had been hoping that the meeting would be organized in the spring of that year).

In fact, Solvay's hasty agreement, without a preliminary examination of the case at hand, is quite surprising... It raises two questions:
- Why did Solvay, a wise and experienced man, react positively to Nernst's "extraordinary proposal" without seeking the opinion of his scientific advisers?
- Why did he take the risk of jeopardizing his credibility by agreeing to sign a letter whose preamble could only be the work of a professional scientist (note that Nernst's name didn't yet appear in the draft of the invitation letter)?

It seems reasonable to assume that Solvay felt inclined to reassure Nernst by agreeing *in principle*, while giving himself some time to study the project in detail (as we will see, Solvay wouldn't give a final answer before March 1911). But this doesn't explain the fact that he immediately accepted Nernst's suggestion to react to a problem which, after all, worried only a handful of physicists.

We shall now see that Solvay had particular reasons to react as he did.

2.4 A glimpse behind the scenes

Let us take a look at Solvay's daily register. By consulting the Notes for April and May 1910, we see[136] that Nernst's proposal couldn't have found a more favourable ground.

This is particularly clear from a note of 16 April, in which Solvay expressed his intention to give a new impetus to the construction of his *Gravito-Materialitic* theory. Among the points to be examined, he mentioned the "big question of specific heats". This is how he described the research program:

> "It will be most important, from the theoretical point of view, to compare, for a given body, the specific heat curves with the corresponding curves of pure and simple specific heats, and also to notice the differences that will appear in the values obtained for different bodies."

By a remarkable coincidence, it was the behaviour of specific heats that preoccupied the Berlin team of chemists. There is little doubt that the subject was raised during the meeting at Goldschmidt's, and that Nernst described the curves which resulted from his measurements. It is equally probable that he mentioned the need to question the theoretical base of the rule of Dulong and Petit[137]: classical kinetic theory.

One imagines Solvay's enthusiasm as he learned of the disarray of this theoretical framework, which in his eyes[138] was nothing more than a *"fanciful theory, incapable of satisfying the philosophical[139] spirit"*.

One also imagines Nernst's relief: not only had he witnessed Solvay's interest in the behaviour of specific heats, he also had acquired the conviction that the industrialist would welcome the idea of becoming the host of an "unprecedented manifestation of higher science".

The positive impressions that Nernst kept of his encounter with Solvay could explain the allusion to the *"wastepaper basket"* in his letter of July 26, as well as his remark *"but you might be interested in the project"*.

As to Solvay, the fact that his mistrust of "generally admitted physical theories" had been amply justified by a leading experimenter from the University of Berlin, could explain his haste in accepting the idea of an International Scientific Council.

Solvay's objectives

Solvay's diary tells us a lot about his personal goals. It reveals his desire to realize the scientific event that has been on his mind for a long time: the publication in Paris of his *"Synthetic Note on the Constitution of the Universe"*.

The need to complete this note before the end of the summer of 1911 had become much more pressing by the prospect of an International Scientific Council. Hence, the industrialist had no other option than to put pressure on his new collaborators in physics, Herzen and Hostelet.

A few words should be said about these two young men.

Born in Florence in 1877, Edouard Herzen (figure 11) was the grandson of the famous Russian writer Alexander Ivanovich Herzen, and the son of Alexander Alexandrovich Herzen, founder of a physiological laboratory at the University of Lausanne[140].

After completing in Lausanne his studies as a mechanical engineer, Edouard defended a thesis in physics in 1901. One year later, he joined the Solvay Company in Brussels and started working in its laboratory.

Hostelet (figure 12) was born in Chimay in 1875. He studied chemistry at the University of Liège and at the ETH in Zurich. Working in Frankfurt with Professor R. Lorenz, he took part in the French edition of the latter's *Elektrochemisches Praktikum*[141].

FIGURE 11: Edouard Herzen in 1921 (standing, centre of the picture), see figure 41.

During a stay in Paris, Hostelet attended the International Congress in Physics of 1900, and collaborated with researchers from Collège de France and from the École Municipale de Physique et de Chimie Industrielles (today the ESPCI). Back in Brussels in 1907, he was hired by the Solvay Company before becoming Ernest Solvay's personal assistant[142].

The current tasks of these scientists were complementary. Herzen had to deal with the original "*Gravito-Materialitic*" approach; Hostelet was in charge of extending the theory to chemistry.

Thanks to Solvay's relentless efforts and the support of his collaborators, a first draft entitled "*Science and the Objective Universe*" was completed on 15 September 1910. Ten days later a new version was released. Herzen was now asked to analyse the crystallographic forms associated with molecules, and to study their energetic contacts. Solvay's aim was to calculate the combination temperatures of chemical elements and of compounds.

FIGURE 12: Georges Hostelet in Cairo (1920's). Courtesy of Mr. Pierre Verhas, Hostelet's grand son.

Further notes were drafted between October 10 and November 9.

Solvay insisted that Hostelet provide him with comparative tables of the properties of chemical elements and associated crystallographic forms, reminding him of the revolutionary character of his *Gravito-Materialitic* approach[143]. He repeatedly asked that the work be completed before the fall of 1911.

The research was organized in a way that hardly varied. Solvay took care of the preliminary calculations. He also kept writing notes. Herzen and Hostelet were in charge of developing the industrialist's ideas, and of rectifying the theoretical points that hadn't been addressed properly.

Precautions taken by Nernst

While Solvay kept busy with the completion of his theory, Nernst sought to define his position in relation to the Council. Determined to stay in the background, he said in a letter to Solvay[144]:

> "*I would like to express the wish that in the invitation letter you do not name me as initiator of the idea of the Council; I would even prefer not to quote me at all, or – when necessary – only in your welcome address...*".

In order to minimize his role in the eyes of the public, he encouraged the industrialist to take an active part in the conference, telling him:

> "*I would also be very happy to learn something about your ideas. At the Council there will certainly be every opportunity to explore them further...*".

An unprecedented project 49

But Solvay knew he wasn't ready to fulfil Nernst's wish. He therefore sent him the following message[145]:

> "*I work independently, in the sense of a search for objective physical bases. My method is deductive and you know its dangers better than I do – you expressed them correctly. I start naturally from a phenomenon which I consider having been rigorously defined. I'm moving forward but haven't yet reached the point where it would be acceptable for me to tell you about my results – still too uncertain. I ask you credit until the moment when it will be necessary to think about the invitations*".

But Nernst's concerns weren't limited to the role played by Solvay. He also needed guarantees about the man he had visited in March 1910, the brilliant physicist on whom the success of the operation depended on...

On November 1, 1910, Einstein received a letter from Emil Fischer[146], Nobel Prize for chemistry in 1902 and co-author with Nernst of the proposal to award Solvay the Leibniz Medal of the Prussian Academy of Sciences.

In this letter, Einstein was told that the boss of a German industry was interested in his work, and that he would donate 15 000 marks to promote his research. The donor's identity wasn't mentioned, but we know[147] that it was Franz Oppenheim, the director of the *Aktiengesellschaft für Anilinfabrikation,* who, with Nernst and Fischer, had been one of the initiators of the Kaiser Wilhelm Institute for Chemistry.

As to the terms of the letter, they leave little doubt about the man responsible for the initiative:

> "*Your brilliant articles in the field of thermodynamics caused a sensation in the world of natural sciences and are frequently commented in our circle, especially since Mr. Nernst undertook to submit your conclusions regarding the law of Dulong and Petit to a check by experiment...*".

2.5 Evolution of the project

The Council's preparatory phase took a new turn during the first months of 1911.

Reassured by the encouraging echoes which arrived from all sides, Nernst decided to give up his "discretion" about quanta. An opportunity presented itself in January 1911. Having been invited to speak at a meeting of the Prussian Academy of Sciences on the occasion of the Kaiser's anniversary, Nernst declared[148]:

> "*Quantum theory is but a singular, even grotesque rule, but which in the hands of Planck with regard to radiation, and between those of Einstein*

as regards molecular mechanics, has produced results so promising that it is the duty of science to submit it to the verdict of experiment...".

Nernst went further: he compared Planck's work to "*Newton's feat for mechanics and to that of Dalton for atoms*". Following Einstein's advice, he added action to words, and resumed his specific heat measurements at even lower temperatures, down to the boiling point of hydrogen. To his surprise, he noticed the appearance of significant deviations from the predictions of Einstein's formula.

Faced with this unexpected situation, Nernst reacted as a shrewd experimenter.

Starting from the existing formula, he modified it in various ways and tested the resulting expressions in order to detect the one that best reproduced the available data. This research, carried out with Lindemann, led to a "*Nernst-Lindemann formula*[149]" with two "Einstein-like" terms. Encouraged by the fact that it reproduced the data at all temperatures, Nernst became convinced that he had improved on Einstein's theory. He therefore embarked on a new research in order to give a physical meaning to the additional Einstein-like term which appeared in the new formula. This endeavour created in him the feeling that he would soon belong to an exclusive circle: that of the true builders of a "successful quantum theory".

Another major event took place in March 1911: Solvay's realization that it was time for him to examine the details of Nernst's proposal for the Council. Thus, in response to Goldschmidt's pressing demands, he sent him this message[150]:

> "*I have been thinking, as promised, about the question of the "Council" since last Sunday, and I find that, under the conditions you have indicated, that is by asking you to take care of all that concerns the convocations, and later of what concerns the meeting (I have only to sign, to approve... and so on), things could be planned for next October... I have a comment regarding the invited members. It seems to me that one could add Marcel Brillouin to the French, and maybe remove an English and an Austrian, if not a German as well*[151]. *See that with Mr. Nernst; I wouldn't like to offend him. You must enable me to defend my choices. I will write to Mr. Nernst a little later on the meaning of what I would like to say at the Council...*".

The addition of a French physicist to the list of invitees seemed important to Solvay, who wanted a balance between the representatives of the great nations. His demand would have major consequences. As we shall see, after several modifications the list of invitees would finally comprise six German, six British and six French members, including Brillouin, Marie Curie and Poincaré.

But why did Solvay mention Marcel Brillouin?

A first element of explanation is that Brillouin (figure 13) was a theoretician, a rare species in France, where physics at the time was largely dominated by experimental research[152]. This fact made him noteworthy to Solvay, who denounced a physics *"immobilized or drowned by experimentation*[153]*"*.

We also know[154] that Solvay showed interest in Brillouin's work. In particular, he may have been struck by Brillouin's position in the ongoing debate on the reality of atoms, and by his vigorous reaction to Ostwald's pamphlet *"The Disarray of Contemporary Atomism"*. In this publication, Ostwald asserted that matter was a mere "invention", and that energy was "the only physical reality[155]".

The point is that both papers – Ostwald's pamphlet and Brillouin's reaction – appeared in a magazine that Solvay used to read: *La Revue générale des sciences pures et appliquées*.

FIGURE 13: Marcel Brillouin in 1911 (sitting, right hand side of the picture), see figure 1.

Herzen in action

Following the recommendations made to Goldschmidt, Solvay sought the advice of Herzen, one of his main collaborators in physics. What did he think of Nernst's draft of the invitation letter?

Herzen, as we shall see, had some knowledge of quantum theory and of the merits of its architects. He also knew of the surprising similarity which seemed to exist between Planck's *energy degrees* and certain aspects of Solvay's *Gravito-Materialitic* theory. But the first thing he noticed when reading Nernst's proposal was the absence in the invitation letter of the names of invited scholars.

Nernst had simply forgotten to mention the list of Council members, a crucial piece of information on the basis of which each invitee would decide either to accept the invitation, or to decline it... Herzen therefore drafted two letters, a revised official invitation and a more personal message to the invitees. He submitted them to Solvay, together with this note[156]:

Dear Mr. Solvay,

You will find enclosed the letter of invitation to the Council in the form that I propose to you. In my mind, you might add to this official invitation a personal letter in which you would present the list of invited scholars, and announce your offer to cover all travel expenses and subsistence costs.

> *Since the Council should last a week in your mind, you could allocate, for example, 500 francs[157] to each member.*
> *Here is the provisional list of invitees to the Council:*
> *Chairman: Planck (Germany)*
> *Secretary: Goldschmidt (Belgium)*
> *Members:*
> *In first line: Lord Rayleigh, Larmor, J. J. Thomson, Rutherford (England); Nernst, W. Wien, Röntgen (Germany); Langevin, Perrin, Brillouin (France); Einstein (Austria); van der Waals, Lorentz (Holland); Knudsen (Denmark).*
> *In second line: Jeans, Schuster (England); Seeliger (Germany); Hasenöhrl (Austria).*
> *The scholars in second line are of a lesser importance, and might be discarded if necessary.*
> *It is up to Mr. Nernst to propose you as honorary president.*

Herzen's draft of the official invitation didn't differ much from the one proposed by Nernst. The main novelties were the replacement of the French term *Concile*, i.e., a Council of cardinals, by *Conseil*, i.e., just a Council, and the reformulation of the preamble, in order to align it with the second letter: Solvay's personal message to his invitees. This was Herzen's proposal for this second letter:

> *Very honoured Sir,*
>
> *We clearly find ourselves in our time at a turning point in the development of our theoretical beliefs about the kinetic and molecular theories. I have been thinking that a Council, bringing together exclusively researchers who are closely affected by this problem and who have a keen interest in it, could favourably influence the development of physics and chemistry. This is why I invite you to participate in a Council which will be held in Brussels, for about a week at the end of next October, and which would be composed as follows...* (enumeration of the names proposed by Herzen *"in first and second line"*).
> *To allow all guests to participate, I am offering an indemnity of 1000 marks (1250 francs[158]) for travel expenses.*
> *I hope to be able to count on your collaboration and express my high consideration.*
>
> *Ernest Solvay*

It should be noted that Herzen had taken care to inquire about the situation of each invited scientist. This enabled him to apply a rule according to which the scholar's country of affiliation had to be taken into account, and not his nationality.

Thus, Einstein, which had been listed by Nernst as a Swiss member, was now presented as an Austrian member, following his recent appointment at the German University of Prague.

As to Herzen's proposal to entrust the Council's presidency to Planck, rather than to Lord Rayleigh, it can be attributed to his awareness of the role that quantum theory would play in the debates (see in section 2.7, his note of October 24 "*Aim and scope of the next Scientific Council in Brussels*", written at Solvay's request).

The Council's presidency

Following Solvay's agreement, Herzen submitted his proposals to Nernst, who probably informed Planck of the idea to entrust him with the Council's presidency.

We have no record of the latter's reaction, but it seems clear that he rejected the idea. Planck had every reason to remain discrete. Not only did he feel "branded" by the failure of his Nobel candidacy, but he may also have been aware of the need to avoid a German presidency.

Last, but not least, the author of the quantum theory had been forced to react to a criticism formulated by Lorentz in October 1910. During his lectures in Göttingen, the father of the electron theory had drawn his attention to the fact that the absorption of radiant energy by quanta was in contradiction with classical electrodynamics, a theory which, as we know, Planck wanted to preserve[159] at all costs.

Confronted with Lorentz's well-founded criticism, Planck had no choice but to give up his initial quantum hypothesis. He soon embarked on the construction of a new "*emission* theory" in which only the emission of radiation would remain quantized.

Hence, being totally absorbed by this recent development, Planck could not do better than to suggest that the presidency be entrusted to Lorentz, a man he admired, who mastered the three Council languages, and who benefitted from the advantage of not being German.

Planck's suggestion suited Nernst for an unexpected reason. It offered him a pretext to get in touch with Lorentz for a very different purpose.

The Kohnstamm-Ornstein case

The event to which Nernst needed to react was the publication in the "Proceedings of the Amsterdam Academy of Sciences" of a paper in which two disciples of van der Waals and of Lorentz, Philip Kohnstamm and Leonard Ornstein, had the audacity to question the validity of the heat theorem. Deeply irritated by the authors' totally unjustified criticism, Nernst embarked on a long correspondence with Lorentz in the hope of drawing him into his controversy with "Messrs. K & O". His anger was such that he didn't refrain from addressing in the same letter questions about the Council and his personal grievances against the Dutch physicists.

The correspondence with Lorentz began on May 2, 1911, with a letter in which Nernst complained about K & O's *"purely polemic"* paper, claiming that it was based on a misunderstanding. The authors had made the error of relying on the van der Waals equation, admitting its validity at the lowest temperatures. Having failed to recover the heat theorem, they simply declared it *"unfounded"*.

Profoundly shocked by K & O's extreme lightness, Nernst told Lorentz:

"I have long since indicated that my theorem is most closely related to Planck's and Einstein's energy quanta. This means, of course, that van der Waals' formula, which is known to be unusable at low temperatures, cannot agree with the heat theorem. I have just answered your Academy, in a strong but appropriate manner, taking the liberty to indicate in the cover letter that I am quite prepared to lower my criticisms in the event that your opinion on the (disputed) *paper would be higher than mine..."*.

Nernst attached some articles on quantum theory to his letter, and reminded Lorentz that he intended to go to Amsterdam, hinting at the possibility of a meeting in Leiden.

Here is a summary of Nernst's later letters to Lorentz (Lorentz's answers have unfortunately been lost):
- May 3, 1911: Second letter. Having learned that Lorentz and van der Waals were the Academy members who had submitted the unfortunate paper, Nernst now asked Lorentz to transmit his reaction to the *Koninklijke Nederlandse Akademie van de Wetenschappen* (KNAW). Aware of the strong words in his response: *"However, I cannot accept the inadmissible tone of Messrs. K & O"*, he indicated that his opinion was shared by Planck. *"It is clear*, wrote Nernst, *that K & O's logic would lead, if adopted, to the outright rejection of quantum theory"*.
- Middle of May, 1911: Third letter in which Nernst presented a softened version of the reaction he intended to submit to the Academy, through Lorentz. The letter contained the outline of a project for the Council, with a provisional list of guests, similar to the one proposed by Herzen (but in which Rayleigh's name had been forgotten and Sommerfeld's had been added). Nernst ended his letter expressing the hope that Lorentz would agree to chair the Council, and that he would accept to present a report on the Rayleigh-Jeans radiation law.
- May 25, 1911: Fourth letter, from which we learn that Lorentz accepted Nernst's demands related to the Council. The author informed Lorentz of his intention to make further suggestions about the list of invitees, declaring that he expected 100% positive answers[160]... Once more, Nernst raised the K & O affair, thanking Lorentz for his clarifying comments on some aspects of the heat theorem[161].

— June 30, 1911: Nernst pointed out that his reaction to K & O's paper had been presented to the Academy by Lorentz and Zeeman, and not by Lorentz and van der Waals, in spite of the fact that van der Waals' name appeared on the document. Surprised at this change of heart, Nernst expressed his disappointment, regretting that van der Waals had chosen *"not to join the presentation"*.

The above letters shed light on Nernst's eagerness to take a stand against any detractor of his theorem... They also testify to his confidence in the Council's results, and to his wish to meet van der Waals in Brussels as one of Solvay's guests.

Invited members and rapporteurs

A letter from Nernst to Goldschmidt, dated May 31, 1911, tells us that the list of Council members had been revisited and completed. One addition had been made on Lorentz's request, who wished to invite[162] his colleague Kamerlingh Onnes (figure 14). The reason was the recent discovery in his laboratory of an intriguing quantum-like phenomenon: the spectacular collapse of the electrical resistance of certain metals at a temperature above absolute zero[163].

FIGURE 14: Heike Kamerlingh Onnes in 1911 (next to Einstein), see figure 1.

Nernst's list indicated that some recommendations, presumably due to Planck and to Solvay (or to Herzen) had been taken into account:
- Lorentz, instead of Planck, was entrusted with the Council's presidency.
- Röntgen and Seeliger had been replaced by Warburg and Rubens, two specialists in the study of black-body radiation.

Three French invitees had been added to those mentioned by Herzen: Marie Curie, Poincaré and Maurice de Broglie, the latter being invited as scientific secretary. This meant that the equilibrium desired by Solvay had finally been established. France, Germany and England were equal: they were represented by six scientists.

Nernst also mentioned two additional Council reports: one by Perrin on "molecular reality", the other by Sommerfeld on a "quantum of action theory".

The Council's agenda now comprised eight subjects[164]:
1. *Deduction of Rayleigh's radiation formula* (Lorentz).
2. *Comparison of the kinetic theory of ideal gases with experiment* (Knudsen).

3. *Application of the kinetic theory to emulsions* (Perrin).
4. *The kinetic theory of specific heats according to Clausius, Maxwell and Boltzmann* (Jeans).
5. *The radiation formula and the theory of action and energy degrees* (Planck).
6. *Specific heats and the "theory of degrees"* (Einstein).
7. *Application of the "theory of degrees" to a number of physical problems* (Sommerfeld).
8. *Application of the "theory of degrees" to a number of physicochemical and chemical problems* (Nernst).

Revealing detail: Nernst had taken care to present the final report.

It had also been decided that the meeting would take place from 30 October to 4 November, and that the indemnity of 1 250 francs for travel expenses would be reduced to 1 000 francs.

A sudden obstacle: Goldschmidt's trip to Africa

While everything seemed to be on track, Nernst found himself confronted with a major problem: Goldschmidt's unexpected departure for the Congo.

The news was a thunderclap. Nernst was greatly annoyed, but he couldn't oppose a trip that was part of a royal mission (Goldschmidt had been appointed by the King to install wireless telegraphy in the colony).

The consequence of this development was most embarrassing: Nernst would have to take care himself of the Council's correspondence. Seeking to save time, he sent this urgent message[165] to Goldschmidt:

> "*Unfortunately, it is not possible for me to have the invitation letters ready for signing by now, if only because my knowledge of French is insufficient. But things have been clarified enough and the invitations could be sent right away... I am enclosing the French text of the invitation, but please send it back to me on the first occasion. So that the letter goes away tonight, I won't let it recopy once more, hoping that you will have it tomorrow morning...*".

Nernst also felt compelled to propose a measure he would have liked to avoid: the addition in the invitation letters of the following sentence: "*Requests and answers should be addressed to Prof. Dr. W. Nernst...*"

The pill was bitter. Contrary to his intentions, Nernst wouldn't be able to hide his role in the launching of the Council. The only thing he could do was to ensure that the news of this extraordinary meeting wouldn't spread beyond the small circle of guests. He therefore asked that the invitations be made on a confidential basis, a precaution all the more necessary when the success of the Council wasn't guaranteed. Confidentiality was also a way of reassuring the signatory of the invitation letters...

An unprecedented project 57

In this regard, we should say that Godschmidt's departure was also a setback for Solvay, who would have to tell Nernst that he could count on the assistance of Herzen and Hostelet, depriving himself of their services when he most needed them. These are his words in a Note[166] of June 1, 1911, on the "gravity of energy":

> "*I count on Herzen to check everything because time is running out; Goldschmidt is about to leave for the Congo and the Council must be convened...*".

The official convocation

On June 3, Herzen went to see Goldschmidt. This is his message to Solvay, after the visit[167]:

> "*I found Mr. Goldschmidt in full preparations to leave for Congo. At a triple gallop, I translated Mr. Nernst's letter and sent you everything so that you could read it. You could then send it to Mr. Tassel... Tuesday morning, I could agree with Mr. Tassel to put us in direct contact with Mr. Nernst*".

The fact that Herzen mentions Emile Tassel is quite interesting. It tells us that Solvay's earliest advisor for physics was still involved in the Council's preparation at the time of Goldschmidt's departure (we will see in section 2.9 that this would no longer be the case at the time of the Council's meeting).

Having been duly informed by Herzen, Solvay sent this message[168] to Nernst (letter of June 5, 1911):

> "*Goldschmidt communicated to me your letter and all that has been planned for the Council. I agree and tomorrow we will take care of the invitations. I hope I can send you one tomorrow, Wednesday evening, and when upon receipt you could send me a telegram, telling me that you agree, everything could be launched before the end of the week (...). It's too bad that Goldschmidt is leaving for Congo. However, in his absence, Mr. Edouard Herzen, who has a lot of freedom, and who is aware of the question, could easily replace him. I keep him at your disposal, and I will ask him to see you in Berlin if you wish (...). I will include your address in the invitation letters for any information that may be requested. If necessary, Mr. Herzen will help you from here or by going to see you...*".

Herzen's proposals, the official summons and Solvay's personal letter, were combined into a single invitation letter. It contained the list of guests and the eight topics to be discussed (Solvay also took care to specify his role). Thanks to Nernst's agreement, the letters were sent to the invitees on June 9, 1911.

The dice had been thrown... Solvay was exhausted, as can be seen in a Note of his personal diary, dated May 17, 1911:

> "*I will think about all this, but I finish this note because Herzen must be impatient. I really feel tired, given that it has been just a year since I got back to this question without letting it go a single moment... It's time I finished my fundamental Gravito-Materialitic theory, I am working hard on it with the Council in mind, for my own edification...*".

Curiously, Solvay didn't mention the other cause of his fatigue: the concern following a call, made to him two months earlier by Wilhelm Ostwald, holder of the 1909 Nobel Prize in Chemistry.

2.6 Ostwald's project for chemistry

FIGURE 15: Wilhelm Ostwald in 1913, see figure 36.

Ostwald (figure 15) had come to Brussels on April 28, 1911. He had turned to Solvay, asking him to create an International Institute of Chemistry. His visit had no relation whatsoever with the Council on physics (Nernst's relationship with Ostwald, his former mentor, had cooled down: the two men had not communicated since 1908, a situation which would continue[169] until after the Great War).

Ostwald's call on Solvay was linked to an event that had taken place in May 1910, on the sidelines of the Brussels Exhibition: the first "World Congress of International Associations".

The Ostwald–Solvay relationship

Ostwald and Solvay were fervent internationalists. They had met on several occasions, notably in 1909 at the University of Geneva, which awarded them an honorary doctorate in physical sciences[170].

Ostwald, one of the founders of modern physical chemistry, was known as a pioneer of "energetics", a current of thought that had caught Solvay's attention. The Belgian industrialist had always been fascinated by the concept of energy. He considered human society as an organism "*subject to the laws of physics and chemistry*".

Having read Solvay's writings, Ostwald granted him in 1907 the title[171] of "*founder of social energetics*". He also took an active part in the industrialist's appointment as honorary member of the Deutsche Chemische Gesellschaft[172].

When in December 1909, Ostwald saw his work crowned by the Stockholm Academy, he had already put an end to his research career. His new priority was the rational organization of chemistry.

Ostwald's commitment was reinforced by the conclusions of the "World Congress of International Associations" which took place in Brussels from April to September 1910. This Congress, chaired by Auguste Beernaert, Belgian Minister of State and Nobel Peace Prize winner in 1909, had been organized by Paul Otlet, a nephew of Professor Héger, whom we have already met.

Otlet, the other Belgian visionary

Otlet's dream was comparable by its audacity to that of Solvay. His aim was to create a universal synthesis of human knowledge. In order to reach his goal he embarked on an extraordinary project: the drawing up of an inventory of all that had been written in the world on all kinds of subjects – a precursor of the World Wide Web.

Otlet's passion for bibliography brought him closer to Solvay, his *"friend from before the first hour"*. Thanks to the latter's help he founded in 1895 the "International Institute of Bibliography".

During his investigations, Otlet had been struck by the fact that 42 international associations had their headquarters in Belgium[173]. This was for him an indisputable sign that Brussels was destined to become the world capital of internationalism. To give substance to this prophecy, Otlet created in 1907 the "Central Office of International Associations". Solvay, helpful as always, supported this new organization. In 1908, he became its president.

Two years later, Otlet took advantage of the Brussels Exhibition: he convened the first World Congress of International Associations. This Congress had no less than six vice-presidents. Among them: Ernest Solvay and Prince Roland Bonaparte (Solvay was also part of the Organizing Committee). International Committees were formed in order to continue the work initiated by the Congress[174].

Ostwald chaired a committee in charge of unifying the systems of units. Solvay became vice-president of a committee for "publication and documentation".

The International Association of Chemical Societies

Encouraged by the World Congress' success, Ostwald conceived an ambitious plan: the creation of an "International Office of Chemical Sciences". A first milestone was laid on April 25, 1911, when Ostwald went to Paris to discuss the matter with Albin Haller, president of France's Chemical Society, and with Sir William Ramsay, president of the London Chemical Society.

These three representatives of German, French and British Chemical Societies agreed to form the nucleus of an "International Association of Chemical Societies (IACS)". Other Chemical Societies were invited to join. The idea was to endow the Association with a rotating presidency so that the annual assemblies would take place in various countries.

The Association's objectives were multiple: standardization of chemical data, constitution of an inventory of the literature for each chemical sector, determination of atomic weights and creation of a universal language specific to chemistry[175].

Ostwald was appointed president of the Association for the first year.

Ostwald's grand design

Still in Paris, the newly elected president realized that he had neither funds nor premises, and that the IACS couldn't act effectively without a permanent "Central Office". Shouldn't he ask Solvay to help him fill these glaring gaps?

Ostwald rushed to Brussels to meet his partner of the World Congress, to inform him of the birth of the IACS, and to present his idea of an "International Institute for Chemistry" that would serve as the Association's Central Office. This unprecedented scientific body would be the "crown of chemistry": it would centralize all its activities.

Aware of Solvay's eagerness to provide his country with scientific institutions, Ostwald explained that nothing had been decided about the location of the planned Institute, but he hinted that a substantial support from his part would be a decisive factor in favour of Brussels... Solvay declared himself in favour of the project, but he asked for details, in particular with regard to the role that the IACS was supposed to play. This positive but prudent reaction, in contrast with his hasty acceptance of the Council project, did not come as a total surprise... The idea of a Solvay foundation for chemistry was not new. A proposal for the creation of a Belgian Electrochemical Institute had been made the previous year by Octave Dony-Hénault, one of Solvay's former assistants[176] at the Institute of Physiology.

Dony had acted on the advice of Frederick Donnan, his British colleague from Liverpool, who directed the "Muspratt Institute[177] of Electrochemistry and Physical Chemistry". Donnan's letter, suggesting the creation by Solvay of a similar Institute in Brussels, has unfortunately been lost, but we can rely on Dony's answer[178] of January 5, 1910:

Highly esteemed Colleague,

I still have to thank you for your very kind letter, and for your wish, in the interest of our country, to see Mr. Solvay crown his career with a physicochemical and electrochemical foundation.

Mr. Solvay is at the moment in Switzerland where he recovers from an indisposition.

I will communicate your letter to him when he returns to Brussels, and I have no doubt that coming from a chemical physicist such as you, it will favourably impress Mr. Solvay. Your wish is, moreover, shared by several of our compatriots, and I am convinced that the support of foreign voices, as authoritative and disinterested as yours, is likely to favour its realization (...).

Perhaps I will attract your interest by giving you a brief view of our scientific situation. We are, as you say, a rich country of intensive industrial production, and our prosperity should be real. It is unfortunately not the case, because our university regime is gangrened by politics that divides, demoralizes and discourages.

In fact, Mr. Solvay has already founded an Institute of Physiology, to which I am attached although not being a physiologist, an Institute of Sociology and a Higher School of Commerce. The first two are assigned to research and teaching, the third to teaching only. Unfortunately, these foundations, much too luxurious in my opinion, have no effect on the development of physical chemistry and electrochemistry.

The School of Mines in Mons is the only one in Belgium which delivers a regular course in industrial electrochemistry. I am the chair holder, but the insufficiency of professorial salaries doesn't allow me to devote myself exclusively to this teaching.

The eagerness of engineering students to enter a physicochemical research career is extremely limited, and it is with great difficulty that I sometimes find a researcher to direct (...). Belgian universities have given little development to physicochemical research. In general, our university colleagues are overwhelmed by the weight of extensive oral classes, numerous exams and rigid programs...

Hence, our overall production is much lower, in quantity at least, than that of our northern neighbours, the Dutch, less numerous and less rich than us.

It is therefore indisputable that a physicochemical and electrochemical Institute would take up an empty place in our country, which needs to be filled.

Would there be any indiscretion to ask you how the control of the founders is ensured at the Muspratt laboratory? How the maintenance fund is managed, how the workers are encouraged, etc.?

It goes without saying that I only ask you to communicate general clauses and regulations that are not of a private nature. But I reckon that it would be possible to draw from your general suggestions some useful ideas for a Belgian foundation...

We have no trace of Solvay's reaction, but there are reasons to believe that he showed interest. We know that the Solvay Company intervened in 1896 to support the creation of an electrochemical Institute in the city of Nancy[179]... Whatever his precise feelings about the Donnan-Dony proposal, it is clear that Solvay couldn't remain indifferent to the idea of an "International Institute of Chemistry" located in Brussels.

Unfortunately, Ostwald's demand came at an inappropriate moment. His partner at the World Congress was totally taken up by his research, and by the Council's preparations... Caught off guard, he asked Dony to get in touch with Ostwald and to gather details about the planned Institute.

Ostwald responded to Dony's letter on May 4, 1911. He sent him a sketch of his great design[180], entitled *"Organization of the International Institute of Chemistry"*.

Dony made clear that it was up to the founder to exercise control over his foundation. He also reminded Ostwald of Solvay's need to obtain guarantees about the IACS.

Did it support Ostwald's plan, and would it accept the idea of having its Central office located in Brussels?

Ostwald replied that he hadn't raised these questions so far, but that he would do so at the Association's next meeting. He declared himself ready to give up the Institute's presidency, and announced his intention to equip its future library with part of his private collections.

A promise subject to conditions

A new development intervened in June 1911. As he prepared to leave for Switzerland for vacation, Solvay realized that he hadn't answered Ostwald's expectations. Seeking to reassure his Germano-Baltic friend, who had only been in contact with Dony, he sent him this personal message[181]:

"I'm leaving on Monday for Pontresina (Kronenhof) in Upper Engadine, for about six weeks, and I wish to tell you that I am ready to devote 250,000 francs to the Institute if the general conditions that I put to its foundation are met. I don't think I would accept any personal activity in the Institute...".

Ostwald thanked Solvay, regretting his decision not to take active part in a project *"which of all his creations could be the richest in consequences for science"*.

Touched by Ostwald's words and insistence, Solvay sent him a new message[182] on July 7, 1911:

"The offer you made me to found an International Institute of Chemistry in my country, to me, the internationalist, the Belgian and the man who has dealt with some success with applied chemistry, has been to my heart. That is why I will do all I can to help you to realize your plan...".

An unprecedented project 63

However, being bound by the confidential nature of the Council, Solvay remained silent about it. He ended his message with this general statement:

"So, I can possibly put a sum, and men, at your disposal, but I cannot promise you an active personal engagement: I am not free enough for that, at least for the moment, the directions in which I am already committed taking precedence over new ones. Forward anyway. I'll be with you as much as I can... Do believe in my gratitude and in my dedication".

Ostwald didn't hide his disappointment. Six days later, he sent a postcard[183] to Dony with the following words:

Very honoured Colleague,

By the decision of Mr. Solvay to give a quarter of a million, but not to collaborate personally, our business is somewhat delayed, because I need to get more money.

My intention is to create a special Committee for this purpose...

Dony didn't react immediately. He waited until October 4, 1911, before transmitting Ostwald's postcard to Charles Lefébure, Solvay's personal secretary, accompanied by these few ironic words[184]:

"Paris was well worth a mass... 250,000 francs are well worth a postcard."

2.7 Back to the Council

Ostwald wasn't the only scientist in search of an active participation from Solvay. Nernst, who still wished to remain in the background, had every reason to encourage the industrialist to take part in the Council's discussions. His expectations were shared by Lorentz, who declared[185] on July 3, 1911:

"I am sure that this overall work, to which we will be very happy to see you take part yourself, will be of the greatest benefit to all of us."

Nernst returned to the charge on August 26, telling Solvay[186]:

"Your presentation of the theme that you described to me will certainly be of great interest to all of us, and if, as you write to me, you could communicate it beforehand, I would be very obliged to you...".

But Solvay didn't feel able to meet this request: his work was behind schedule.

Moving on to something else, he informed Nernst of his intention to announce the holding of the Council to the King. The chemist reacted with enthusiasm:

> "Regarding your benevolent efforts to spend the evenings during the Congress, a reception by the King would naturally give a very particular sparkle to the event, and would be a precious memory for all participants...".

Solvay sent a letter to King Albert on September 27, 1911, inviting His Majesty to devote some attention[187] to *"this unprecedented manifestation of higher science...".*

The King answered two days later[188], expressing his willingness to *"follow closely this manifestation of higher science...".*

But there would be no reception at the Royal palace... The Council would be eclipsed by unexpected political tensions: a rebound in the Belgian school war (the Schollaert affair), and the repercussions of the "Agadir Coup", a Moroccan crisis[189], the solution of which threatened the security of the Belgian colony.

Solvay's preliminary study

Pressed on all sides, Solvay decided to have a summary of his theory printed for distribution to the Council members. Having done so on October 15, 1911, he sent his preliminary study *"On the establishment of the fundamental Gravito-Materailitic principles"* to Nernst, who acknowledged receipt in a letter[190] to Goldschmidt:

> "I received Mr. Solvay's very interesting study; I will write to him about it tomorrow evening."

In fact, Nernst didn't feel qualified to study Solvay's work. Preferring to leave it to Planck, he entrusted him with the document. Planck read the study, wrote a report and sent it to Nernst, who transmitted it to Solvay.

These are Planck's words:

> "I can say in all sincerity that, from the beginning of my reading, my interest has steadily increased, not only because the author proves he knows the laws of physics, and in particular those of the movement of planets, in a way that would honour a professional theoretical physicist, but especially by the description and by the mode of observation which are absolutely independent and original. It seems to me that one could well characterize this mode of observation by saying that the author defines the "state" of a material point revolving around a fixed point – not as in classical mechanics by means of its behaviour at a given moment – but by considering the entire trajectory from the start, and by studying its essential particulars[191]...".

Planck tempered his praise with some critics, indicating in particular that *"he couldn't immediately agree with Solvay's views, mainly because they did not take sufficient account of electrodynamic phenomena"*. However, he concluded his report with this reassuring note:

"A point worth to be emphasized is the fundamental role that the author assigns to molecular surfaces and to contact actions, role on which he bases his hypothesis of a "direct" and an "inverse" ether. Considered from this point of view, chemical phenomena appear in a new light, and catalysis acquires a general and primordial importance for all kinds of chemical reactions. In any case, I will be happy to receive the finished work, and to study it at greater length...".

We can imagine Solvay's satisfaction when reading Planck's judgment. Thrilled by the positive reaction of this leading physicist, he decided to outline his "*Gravito-Materialitic* theory" in his opening address to the Council.

Furthermore, he asked Herzen to prepare two notes: one on the meeting's agenda, the other on Solvay's personal position with respect to the conference.

The purpose of the first note, dated October 24, 1911, was to inform Solvay about the reports that would be discussed:

Aim and scope of the next Brussels Scientific Council

The reports to the Council can be divided into two groups, the first of which should include presentations on the following subjects:
— *Deduction of Rayleigh's radiction formula* (Lorentz).
— *Comparison of the kinetic theory of ideal gases with experiment* (Knudsen).
— *The kinetic theory of specific heats according to Clausius, Maxwell and Boltzmann* (Jeans).

The goal of these three reports[192] *is to determine which results of the old mechanical design of matter are in agreement with experiment, and which are not.*

The reports of the second group are intended to show how a new theory, much more adapted to experiment than the above one, can be built by replacing the mechanical design of matter by a new one, founded on an electromagnetic basis and on a daring hypothesis due to Mr. Planck:
— *The radiation formula and the "theory of action and energy degrees"* (Planck).
— *Specific heats and the "theory of degrees"* (Einstein).
— *Application of the "theory of degrees" to a number of physical problems* (Sommerfeld).
— *Application of the "theory of degrees" to a number of physicochemical and chemical problems* (Nernst).

> Mr. Planck's hypothesis that the energy of an atomic or molecular system can only vary by successive degrees, depending on the period of vibration, is at the heart of the questions to be dealt with in the Council. This hypothesis, which cannot be deduced from Newtonian mechanics, shakes classical physics in its very foundations: it condemns the application of ordinary mechanics to molecular systems, and rejects the use of differential and integral calculus, without special corrections, which implies perfect continuity in the variations. The theory seems more satisfactory than the old mechanical theory, but is it the only possible one? This is what the future will tell.

The second note, dated October 26, was intended to inform the Council members on the position of their host:

> **Mr. Solvay's position in the Brussels Scientific Council**
>
> From the first half of 1910, Messrs. Nernst and Planck, from Berlin, projected the convening of a Congress that would sanction the need for a reform of the old mechanical theory of matter. Mr. Planck feared at the time that the call wouldn't be heard. However, little by little, the need for this reform became more and more evident. In the meantime, Mr. Nernst, passing through Brussels, had met Mr. Solvay at Mr. Goldschmidt's. The immense interest of Mr. Solvay in any discussion concerning the constitution of matter, and his well-known generosity, put an end to the hesitations of Messrs. Nernst and Planck. It was decided that the Council would take place in Brussels. Mr. Nernst took it upon himself to organize it in view of Mr. Planck's ideas.
>
> Mr. Solvay has always been concerned with the problem of the constitution of matter.
>
> His Gravito-Materialitic theory, published on the occasion of the Council, testifies to this constant concern since 1887, and even (see ref. below[193]) since 1858. Mr. Solvay gives, among other things, a representation in space of gravitational energy, comprising surfaces and volumes of invariable universal atoms. This representation implies a variation of energy by jumps or degrees, instead of a continuous one. This notion of energy degrees appearing in Mr. Solvay's ideas is the only point of contact with Mr. Planck's theory that I have been able to point out at Mr. Solvay's special request[194].
>
> Mr. Solvay intends to stay out of the Council discussions, regarded as too special.

To complete his work, Herzen drew up a biographical note of each Council member, and illustrated it with a photo[195]. Here are Herzen's notes on three key-figures: Lorentz, Planck and Einstein. They are revealing of the way in which the respective merits of these physicists were perceived at the time.

An unprecedented project   67

1. **Mr. Lorentz**, *Dutch member, President of the Council*
Lorentz, Hendrik Antoon, born in Arnhem in 1853, is since many years Professor of Mathematical Physics in Leiden. He published a lot of Memoirs on major problems of electricity, optics and heat, but he is specially known for his admirable theory of electromagnetic phenomena, the application of which he constantly extends with many followers. This powerful spirit is often compared to Poincaré; but while the latter is famous for his sharp criticism, Lorentz is a daring builder. It is his electron theory which in 1896 led Zeeman to the capital discovery that the radiations emitted by a body are modified by the action of a strong magnetic field.

Always guided by the same theoretical views, Lorentz retraced the quantitative laws of emission and absorption, thermoelectric phenomena, etc. His important theory, which makes it possible to group together in the same set so many facts of various origins, is in the process of reviewing the fundamental notions of mechanics. The concept of mass vanishes to merge with that of energy. It is energy which is inert, matter resisting the change of speed in proportion to the energy it contains. The notions of space, time, energy itself, lose their absolute meaning, so that Lorentz's theories force philosophers, like physicists, to recognize the empirical origin of these notions.

One can say, without exaggeration, that this mathematical physicist has renewed the aspect of the problems which torment the metaphysicians.

2. **Mr. Planck**, *German member*
Max Planck, born in 1858, is a Professor of mathematical physics at the University of Berlin. Being extensively trained, and endowed with a quite remarkable mathematical faculty, he has developed his activity in all areas of physics and chemistry where there was material to create mathematical frameworks.

The importance of Planck's contributions appears particularly in his famous "Lessons on thermodynamics", and "Lessons on thermal radiation" - which at the same time reveal his didactic qualities. It is especially in the study of radiation that he gave the mark of his power and originality. His radiation theory, whose foundations date back to 1901, is a grandiose application of "atomistics": an attempt to formulate the universal function which governs the distribution of energy in the normal spectrum.

The way Planck deals with the problem results in the "atomistics" of energy.

A very interesting presentation of his theory can be found in his "Eight lessons in theoretical physics", given in New York in 1909.

3. **Mr. Einstein**, *Austrian member*
Albert Einstein, born in 1879, is a Professor of Theoretical Physics at the University of Prague. He is a brilliant continuator of the works of Lorentz and Planck. We owe him, in particular, a very clear study of the principle of relativity in the concepts of the first, and important contributions to the

theory of radiation, that has been based by the second on the hypothesis of a discontinuous structure of energy. The results of his theoretical works served as a guide for Perrin's fine research on Brownian motion, as well as for the development of Nernst's theory of chemical equilibria.

2.8 The first Council on Physics

The Council took place, as planned, from Monday October 30 until Friday November 3, 1911. The members were welcomed on October 29 at Hotel Metropole, where a room had been made available to them (see figure 1).

Only fly in the ointment: only two of the six English invitees – Rutherford and Jeans – answered the call. Four representatives of the British elite declined the invitation, for various reasons: Lord Rayleigh[196], Sir Joseph John Thomson, Sir Arthur Schuster and Sir Joseph Larmor. Their refusal to come to Brussels may be linked to Solvay's reputation for supporting the policy of social reform in England[197], a policy advocated by the members of the *Eighty Club* (four MP's attached to this *Club* had been invited by Solvay to attend a meeting at the Institute of Sociology, in order to describe the orientation of the English Liberal Party in social matters[198]).

As to Nernst, he came to Brussels accompanied by his assistant Lindemann. This last-minute arrangement was a response to Goldschmidt's request[199] to be supported in his work by a deputy manager. Lorentz, most probably, welcomed the presence of a third "English physicist" as a means of alleviating the absence of four major British invitees.

Another regrettable defection was that of Johannes D. van der Waals, the Dutch Nobel laureate for physics of 1910. Much to his dismay[200], and despite his efforts to convince van der Waals to attend the meeting, Nernst had to face the fact that he wouldn't be able to defend his theorem in the presence of the man whose equation had inspired its recent detractors.

Opening speeches

Solvay spoke first, thanking his guests on his own behalf and on that of Nernst. Then, by way of introduction, he evoked the links that might exist between the subject of the conference and his personal work:

> "In accordance with my letter of convocation, and before you address the agenda of the Council, I would like to say a word about the Gravito-Materialitic study which I had printed on the occasion of our meeting. Each of you received a copy, but too late to be able to read it. You will see, when it will be possible, that the basis of my research is common with that of yours, in the sense that both relate to the constitution of matter, space and energy. And this proves that when Mr. Nernst hadn't first thought of convening a Council on the subject, I might have been tempted to do so,

An unprecedented project 69

by a curious encounter of situations – that is if I had been bold enough to submit my study to you.
I do think with firmness, that it leads to the exact, and therefore definitive, knowledge of the fundamental elements of the active Universe...".

Solvay went on, exposing his working method and his philosophical position:

"The method I followed was deductive. I started off with a general preliminary concept, capable, in my opinion, of satisfying the most scrupulous constructive philosophical spirit. Of course, as you will notice, my fundamental work is far from finished; it is neither perfect nor complete. Many elements, likely to complete those that have already been established, are still missing, and it is with extreme regret that I felt compelled to expose my hastily acquired results (...).
In one year, the study will no doubt have reached an acceptable degree of general completion, and I regret from this point of view that the Council couldn't be adjourned until then. You will see from what precedes that this study is basically of the order of physical philosophy rather than of current physics.
For forty years I have expressed the opinion that in the mental reconstruction of the active Universe, on which we all work with conviction, the last word of supreme enlightenment must be said by the philosopher, rather than by the experimenter. In this sense, it is no longer experiment which will be the source of computation, it is computation which will henceforth provoke experiment (...). This purely experimental route should be abandoned in our time, such is my thought, because the modern philosopher who wants to be precise, correct and curious – that is to say exclusively objective – seeks to perceive the active Universe as it is in reality, and not in the artificial representation of its multiple phenomena, often with great fantasy, as is the case, in my opinion, with the kinetic theory of matter...".

Next, it was Chairman Lorentz who took the floor:

"What will be the outcome of our meetings? I don't dare to predict it, not knowing what surprises could be in store for us. But since it is safe not to rely on surprises, I will admit as very likely that we will contribute very little to immediate progress. Indeed, progress is made by individual efforts, rather than by deliberations in Congresses, or even in Councils, and it is quite possible that, while we are discussing a problem, an isolated thinker in some remote part of the world will find its solution.
Fortunately, there is nothing in this that should discourage us. When we fail to overcome the present difficulties, we will at least be excited and prepared to tackle them again, each in our own way, and we will gain from here ideas and views which will be of great use to us...".

The last speech was that of Nernst. His words weren't those of a theoretical physicist, accustomed to work on his own. He spoke as a team leader, engaged in an ambitious experimental program. According to the Council's designer, the meeting was in line with a historic gathering: the *Karlsruhe Congress* of chemists[201], held in 1860 to clarify the chemical nomenclature and to settle the question of molecular and atomic weights. Resolutely optimistic, Nernst ended his speech by highlighting his choice of a new and promising working method:

"*In this case*, he said, *the chances of success are greater than in Karlsruhe, because the deliberations of this Council will have been informed by eight carefully prepared reports.*"

Council sessions

Contrary to what had been planned, the Council was supposed to hold its sessions at Solvay's Institute of Physiology, most of the meetings took place in a small room of Hotel Métropole. This arrangement seemed convenient to Chairman Lorentz and to his colleagues: it favoured the exchange of ideas in the evening after dinner, or in the morning at breakfast[202].

The lecturers had a whiteboard on an easel and a projection lamp, elements considered sufficient for most of the presentations. However, Nernst and Perrin argued that the small hotel room wasn't suited for the presentation of their reports, encumbered by numerous data tables. It was therefore agreed that both would benefit of the Institute's large amphitheatre, where a special session would be organized on November 1. This decision had an unexpected effect: Nernst, who wanted to speak last, felt compelled to give up his place to Einstein.

The Council proceeded as follows (the proceedings have been reported[203] by Maurice de Broglie, who took notes on the spot):

Monday, October 30
Morning session, hotel Métropole (chair: Poincaré)
– Report by Lorentz: "*Application of the equipartition of energy theorem on radiation*"; discussion.
Afternoon session, hotel Métropole (chair: Lorentz)
– Lindemann's presentation of a *letter* from Lord Rayleigh; discussion.
– Report by Jeans on "*The kinetic theory of specific heats according to Clausius, Maxwell and Boltzmann*"; discussion.

Tuesday, October 31
Morning session, hotel Métropole (chair: Lorentz)
– Report by Warburg: "*Experimental verification of Planck's formula in the region of high frequencies*"; discussion.

- Report by Rubens: "*Verification of Planck's radiation formula in the long wave region*"; discussion.
Afternoon session, hotel Métropole (chair: Lorentz)
- Report by Planck: "*The black-body radiation law and the hypothesis of elementary action quantities*"; discussion.

Wednesday, November 1
Morning session, hotel Métropole (chair: Lorentz)
- Report by Knudsen: "*Kinetic theory and the experimental properties of ideal gases*"; discussion. The session ended at 11:40.
Afternoon session, Institute of Physiology, Leopold Park (chair: Lorentz)
- Report by Perrin: "*Evidence of molecular reality*". The discussion was postponed to the next day, November 2.
- First part of Nernst's report: "*Application of the theory of quanta to a number of physicochemical and chemical problems*".

Thursday, November 2
Morning session, hotel Métropole (chair: Lorentz)
- Discussion of Perrin's report.
- Continuation of Nernst's report; initial discussion. Presentation by Kamerlingh Onnes of electrical resistance curves for various metals.
Afternoon session, hotel Métropole (chair: Lorentz)
- Further discussion of Nernst's report.
- Report by Sommerfeld: "*Application of the element of action theory to non-periodic molecular phenomena*"; discussion.

Friday, November 3
Morning session, hotel Métropole (chair: Lorentz)
- Report by Langevin: "*Kinetic theory of magnetism, the magnetons*"; discussion.
- First part of Einstein's report: "*The current state of the specific heat problem*".
- Private meeting of Solvay with the members of a "Select Committee", chaired by Lorentz.
Afternoon session, hotel Métropole (chair: Lorentz)
- Continuation of Einstein's report; discussion.
- General discussion and closing speeches.

As we see, the Council's program underwent last minute changes: addition of three reports (Langevin, Rubens and Warburg), changes in the title of some reports.

Of the eleven reports presented during the conference, only those of Warburg, Rubens and Langevin had not been made available before the meeting (Nernst only cited eight reports in his opening speech...). Seven reports were in German[204] (Warburg, Rubens, Planck, Knudsen, Nernst,

Sommerfeld, Einstein), three in French (Lorentz, Perrin, Langevin) and one in English[205] (Jeans).

The members were allowed to speak in the language of their choice: French, German or English. President Lorentz took the trouble to translate the questions and answers. He also summarized the debates at the end of each session.

FIGURE 16: Maurice de Broglie in 1911 (standing, right hand side of the picture), see figure 1.

De Broglie (figure 16) took note of the interventions that were made in French; he gathered the handwritten notes of those that had been made in German or in English.

We learn from his report that the rapporteurs were regularly interrupted, in the manner of an informal seminar. We also notice that the discussions did not always take place according to plan. This was notably the case with Perrin's report, the length of which exceeded the allotted time. Deeply irritated by this lack of rigor, Nernst asserted his right to benefit from the amphitheatre for his own presentation. He demanded (and obtained) the postponement of the discussion of Perrin's report.

The proceedings of the Council were to be published in French, out of respect for Mr. Solvay. The writing of the book was entrusted to de Broglie and Langevin, the latter accepting to take care of translations. The official volume was published in 1912 by Gauthier-Villars. Its title "La théorie du rayonnement et les quanta" is indicative of the change that had taken place in the Council's centre of gravity: the questions of molecular theory, highlighted by Nernst, had gradually been overshadowed by the more fundamental problem of radiation, an evolution apparently due to Einstein and Lorentz.

The Council's Minutes

The Notes taken by de Broglie reveal the existence of significative differences between the course of events and the report presented in the Council's official volume.

One such point concerns a "report" attributed to Kamerlingh Onnes, although nothing of the kind had been programmed. In fact, we know thanks to de Broglie that Kamerlingh Onnes intervened at length during the discussion of Nernst's report. We also know that he presented a graph that dramatically illustrated the new phenomenon of superconductivity[206], a phenomenon of quantum appearance that had been observed in his laboratory for the first time in April 1911. Hence, it is more than probable that Lorentz decided to highlight his colleague's discovery by presenting his contribution in the form of a Solvay report. This slight distortion of the truth provides us with an explanation of an anomaly: the absence of any real discussion[207] of this most challenging report.

A more disturbing element is the disappearance from the official report of a crucial statement made by Planck. We know that the author of the emission theory suspected the existence of a fundamental link[208] between the heat theorem and his quantum of action. Having omitted to comment on it in his report, Planck took advantage of a later occasion and declared:

"It seems that an explanation of Mr. Nernst's theorem could be found in the theory of quanta."

Nothing could please Nernst more than this first-hand justification of his claim. Reported by de Broglie, Planck's statement appeared in the Council's provisional Minutes that were sent to the various contributors on December 23, 1911. Yet, no trace of it remained in the official volume, published by Gauthier-Villars.

One can wonder about the reason for this deletion. A possible clue can be found in the fact that Einstein disputed Nernst's claim and that Lorentz followed suit by declaring that he had felt inclined to agree with Planck, but that Einstein's objections had forced him to change his mind. Indeed, we know that the speakers, upon receiving the report of their contribution in the provisional Minutes, were invited[209] to mention possible remarks that they hadn't made during the meeting, and which could still be added to the text in the form of a footnote[210]... We may therefore assume that Planck made use of this provision in an unexpected way, by asking for the deletion of his "unfortunate" statement.

The Council Minutes also tell us about a seemingly innocuous fact, but which would prove decisive: the interruption of Einstein's report on the last day of the conference.

The idea was to allow Solvay to have a "private" meeting with the members of a *Special Committee*, chaired by Lorentz and comprising seven members: Marie Curie, Brillouin, Rutherford, Nernst, Kamerlingh Onnes, Warburg and Knudsen.

We have no trace of the conversations that took place during the meeting, but we presume that Solvay asked his guests to organize new experiments that might confirm his views on the source of the energy at work in two phenomena[211]: Brownian motion and radioactivity. There is little doubt that he accompanied his request with a pledge of financial support. It is also probable that he confidentially communicated his intention to create an International Institute that would grant subsidies to promising researchers of all nationalities, as had been suggested[212] by Goldschmidt.

Impressions of three members

FIGURE 17:
Arnold Sommerfeld in 1911 (centre of the picture), see figure 1.

Here are Sommerfeld's (figure 17) impressions in a letter[213] to his wife on October 31, the day after dinner at Solvay:

"We are installed in a hotel of downright stupendous pretentiousness. Each of us has a bathtub and a toilet in his room. I take a bath every morning. We are Mr. Solvay's guests at all meals, noon and evening. Lunch doesn't include less than five courses. It's crazy.

Mr. Solvay is very sympathetic. He told us of his discoveries with great tact, but such that he himself cut short the discussion of it. There are three Solvay Institutes in Brussels, all built by him and maintained as scientific institutions.

My wardrobe is quite appropriate. Einstein naturally went to dinner with the Solvay family without a tailcoat... He doesn't have that kind of clothes. Lorentz asked me about you, he remembers the names of our children... He performs his task admirably... I slept until 10 a.m. today, and plan to be at the Council meeting at 11 a.m. We meet every day for almost five hours. Yesterday evening I had a Frenchman on my left and an Englishman on my right; I spoke with them in turn...".

This cordial atmosphere, described by a German physicist, takes on its full meaning when we remember that Europe was shaken by the Moroccan crisis, and that tensions between France and Germany were extremely high[214]. An armed conflict between the two nations had been narrowly avoided, thanks to the mediation of England and Russia. We know from Brillouin's later testimony[215], that the affair occupied all minds, giving rise to lively discussions among the Council members.

Most significant are Lorentz's impressions[216], which he shared with his daughter, Geertruida de Haas on November 28, 1911:

"I arrived with Onnes at hotel Métropole on October 29, around 6.30 p.m. We were received royally. A meeting took place at 8.30 p.m. The first session started the next day at 10 a.m. Solvay is a self-made man; he founded with his brother a soda production industry which now extends to the whole world. This great industrialist is an idealist firmly convinced that the progress of science, especially the physical sciences, will make mankind happier. Hence, his previous foundations, three superb Institutes (of physiology, sociology and commerce), and the current invitation.

He is a man of great nobility. We presented in turn our reports (first mine on the old theory of radiation) and discussed them with passion in three languages (the German and the British preferring to express themselves in their language). It was very enriching and everyone was very satisfied, even if none of the issues were resolved – this is not possible when you are twenty.

However, the debates gave us a lot of food for thought. Discussions continued into the evening, but as president I hardly took advantage of them, because I had to meet regularly with the secretaries.

Friday noon we had our last session; it was followed by a short closing meeting. Saturday, I had my last interview with the secretaries; after that I went with Onnes to Verschaffelt's lab, and to Solvay for lunch. Then I said goodbye to Goldschmidt, who had offered us dinner. I barely had time to take the last train home... Solvay also received us for dinner one afternoon.

Madame Curie is tireless and aware of everything; Rutherford friendly and joking; Wien very pleasant and friendly, like everyone else. Einstein is insightful as always, he sees further than all others. Often taking part in the debates, he finds a way each time to contradict his opponents, but in such a charming way that no one would dream of taking offense.

I had for myself a magnificent room, with toilet and bathroom, near the meeting room (we had our meetings at the hotel, where we were Solvay's guests). I could also withdraw at every break (of a quarter of an hour) and settle down comfortably to refresh my ideas. After my return I repented having taken so much part in the discussions, because I still had to put my remarks in writing and send them to the secretaries...".

And here are Brillouin's impressions, part of a paper[217] written a few years later in homage to Lorentz:

"What a wonder to hear a stranger speak our language which such perfection of form and substance (...). Mr. Lorentz speaks a language which, by its simplicity and ease of syntax, by the exact property of the form, invincibly reminds us of the best writers of the 18th century. And what delicacy of feelings, what a fine expression of nuances. Above all, what an emanation of benevolence and sympathy... This impression has only grown and strengthened in those who, like me, had the privilege of seeing Mr. Lorentz at work as the organizer of the Brussels first Solvay meeting in 1911... Knowledge and practice of the English and German languages allow Mr. Lorentz to grasp, in all its folds, the thought of the interlocutors, and to respond to it with the most nuanced precision. And he seems tireless, despite switching from one language to another, in the discussion of theories and hypotheses that are so far away from classical notions.

During these long and laborious meetings of 1911, in the small slightly overheated room of hotel Métropole in Brussels, Mr. Jeans and Mr. Rutherford used to leave us at teatime to take a few moments of rest. The others, at six o'clock, felt the need to get some fresh air on Boulevard Anspach, to give the mind a little rest, or to talk head-to-head about the difficulties raised during the meeting. Mr. Lorentz, by a sudden change of concern, had to rush to Mr. Solvay, to develop the financial program by which the latter had decided to prolong the fascinating meeting of 1911...".

Closing speeches

Five speeches were made on Friday 3 November. We have chosen to reproduce two: Solvay's, which appeared in the official volume[218], and Poincaré's (which wasn't reproduced, probably on Lorentz's request, for the sake of modesty).

Poincaré
Reported by de Broglie, Poincaré's words were apparently his last before an international audience:

Madam, Gentlemen,

I believe to be the interpreter of all our colleagues in addressing all our thanks to our president, Mr. Lorentz. We admired the perfect tact with which he directed our discussions, the elegance with which he handles the three languages, as if he had three mother tongues, or even four – I forgot Dutch.

We admired the luminous clarity with which he summed up our debates at the end of each session, and with which he just summarized all of our work.

We also admired the benevolent impartiality with which, by giving an account of opposing and almost contradictory opinions, he managed to bring out the plausible elements that each of them contained. The job wasn't easy, the difficulties he encountered being not of the kind faced by presidents of parliamentary assemblies – we were not ferocious beasts ready to devour each other – but to be of another nature didn't make them less great.

We are sailing on a completely unknown sea, and we needed an experienced captain.

If we hadn't found him we would probably have lost our way.

I should also remind you that if we raised many ideas that seem to open up a whole world to us to conquer, the origin of all these ideas can be found in Mr. Lorentz's classical memoirs. This fact alone provides us with one more reason for expressing our gratitude to our president.

Solvay

Madam, Gentlemen,

I thank you more warmly today than I did on Monday at the opening session of the Council, now that I have seen you at work, and that I have been able to appreciate the enormous amount of labour that you have provided without truce, rest, or distraction. I am deeply moved, just as I am to have been able to observe the great superiority of the presidency of our eminent Mr. Lorentz.

You will have fixed the current state of physical science in one of its fundamental directions, in meetings which will occupy a remarkable place in history, I have no doubt. The printed reports and discussions will constitute a monument that centuries will respect.

But despite this, Gentlemen, and despite the beautiful results that have been obtained thanks to the Council, you will not have resolved the difficulties of the present hour, nor above all indicated the frank and clear path which leads to the exact, calculated determination of the main simple elements, in simple operation, that the philosopher sees as constituting the active universe, also simple in its entirety, towards which my personal study is directed. Thus, my duty obliges me to tell you that I keep intact the convictions that I expressed at the opening of the Council.

If nothing stands in the way, and according to a general wish which exists, I think, we will be able to meet again in 1913, and I will be happy to invite you once more.

And then, Gentlemen, I hope to be able to defend my Gravito-Materialitic thesis in parallel to your theses, expecting that by then my study will have acquired the required completion for this purpose. I acted all the time so that it couldn't influence your previously fixed deliberations, but also so that it would be recorded on the occasion of the Council's meeting.

In the meantime, if I might formulate a wish, it would be to follow-up the experiments I have in mind in my search for the origin of the energy which manifests itself in Brownian motion and in radioactivity, being still firmly convinced that it comes from the external medium, and not from the Brownian medium or from the radioactive bodies themselves.

You, specialists, could easily, I believe, help me to obtain satisfaction on this matter, by resolving the question, one way or the other, under my control. I apologize for my tenacity in this regard. Ask yourself, Gentlemen, if it isn't worth, or doesn't prolong, the industrial stubbornness of my youth, which I would have been wrong not to have – you are perhaps aware of it – even more today than before, after following your discussions (...).

I have no doubt that you will accept, Madam, Gentlemen, to do me this pleasure that I ask of your benevolence; I will help you with all my power.

2.9 Council results

In the opinion of several members, the meeting hadn't produced the expected answers. As pointed out by Solvay, no substantive issue had been resolved.

Among the disappointed participants, there was the Council's initiator. Despite his optimism, Nernst hadn't obtained what he most hoped for: the acceptance of his idea that his theorem could be deduced from Einstein's molecular quantum theory. His bitterness was all the greater as the disavowal had been pronounced by his champion.

Indeed, Einstein didn't deliver the expected reassuring message. Rather than insisting on the victory of Planck's hypothesis in the molecular domain, he explained that in its present form quantum theory couldn't explain important facts, such as the behaviour of thermal conductivity of matter in the vicinity of absolute zero[219].

As to the Nernst–Lindemann formula, presented in Nernst's report as an improvement on Einstein's result for the specific heats of solids, it failed to win general support. Even worse, Nernst's daring speculations about the theoretical significance of the new formula didn't convince Einstein, nor most of his colleagues.

Regarding radiation, the situation was just as grim. Lorentz opened the debate by recalling that Newtonian dynamics and energy equipartition led to a black-body law in total disagreement with experiment. He concluded that the only way out was to admit the existence of actions foreign to classical mechanics, but that he had no idea of their nature, nor of the way in which they should be taken into account.

Speaking about a possible "structure" of radiation, Lorentz reaffirmed his opposition to the existence of light quanta, indicating that he couldn't accept the idea of radiant energy concentrated in regions of such a small size.

From Planck's point of view, the outcome of the conference wasn't more brilliant.

His attempt to reconcile his radiation law with classical electrodynamics by means of a new quantum theory, the *emission theory*, didn't provoke frank adhesion on the part of the Council members.

Planck certainly scored a point by showing that his radiation law could be derived in various ways. But he had to admit that his emission hypothesis implied a fundamental asymmetry between the *quantized* process of emission of radiation, and its *continuous* absorption[220]. On the other hand, the new hypothesis produced a change in the energy of Planck's oscillators, with the appearance of a term independent of temperature. This prediction of a *"zero-point energy"* opened the way to various speculations.

All in all, Planck's presentation of his second theory gave rise to a lot of criticism. Einstein denounced the fact that Planck persisted in using probabilities, without defining them from a physical point of view. Poincaré observed that neither of the two theories could be extended to systems with several degrees of freedom. Sommerfeld signalled contradictions between the predictions of the new theory and certain X-ray data.

Furthermore, Lorentz and Poincaré criticized Planck's idea that the quantum hypothesis should no longer be viewed as an "energy hypothesis", but as an "action hypothesis".

The other theoretical novelty, Sommerfeld's "*h-hypothesis*", also produced lively discussions. Everyone welcomed a theory that, in contrast with Planck's, could be applied to a variety of non-periodic phenomena, such as the production of X-rays, the emission of gamma rays and of photoelectrons. Unfortunately, Sommerfeld's calculations remained largely inconclusive. His results were discussed at length, notably by Marie Curie and by Rutherford who questioned their significance. Problems of a more theoretical nature were pointed out by Einstein and by Poincaré.

In addition to their mixed feelings about the new theoretical proposals, the Council members had to face Warburg's reservations about the validity of Planck's law. Commenting on the recent attempts at verifying its predictions, the President of the Physikalisch-Technische Reichsanstalt (PTR) indicated that the confidence in the data of 1900 had been called into question in the light of the latest results. His conclusion was the following: "*The researches conducted so far do not disprove Planck's formula, neither do they provide a complete verification*".

Einstein, the man who "*saw further than everyone else*", extended his report beyond the specific heat problem. Aware of the shortcomings of Maxwell's electrodynamic theory, he recalled his statistical results which pleaded in favour of a discontinuous structure of radiation. However, he acknowledged the failure of his relentless attempts to construct a new electromagnetic theory, capable of accounting for the contradictory aspects of light. Expressing deep concern at the situation, he declared[221]: "*Anyway, it appears from the above considerations that our electromagnetism cannot, any more than our mechanics, be put in agreement with the facts*[222]".

Einstein then focused on the problem of black-body radiation. Referring first to Planck's law, he ended his discussion with a distressing comment[223]:

"*I had no other goal here than to show how fundamental are the difficulties in which the radiation formula leads us, even if we consider it as a simple experimental fact.*"

Then, speaking about Planck's justification of the law, by means of his quantum hypothesis, he pointed at the magnitude[224] of the challenge:

"*These discontinuities which make it so difficult to accept Planck's theory really seem to exist in nature The difficulties that a satisfactory theory of these fundamental phenomena must overcome seem, at this moment, insurmountable. Why does an electron in a metal struck by Röntgen rays take the high kinetic energy observed in the secondary cathode rays?*

All the metal is in the field of the Röntgen rays; why is it only a small part of the electrons that pick up the speed of these cathode rays? How come that the energy is absorbed in extraordinarily few points? How do these few points differ from others? We have no answers to these questions and to many others...".

More than anyone else, Einstein was aware of the depth of the crisis. In a letter to Michele Besso[225] of December 26, 1911, he confessed that he hadn't learned anything in Brussels that he didn't already know. He even told his friend that the Council made him think of the *"lamentations over the ruins of Jerusalem"*.

In fact, Einstein's disenchantment about quanta went back to the spring of 1911, when he wrote, again to Besso[226]:

"I no longer ask whether these quanta really exist. Nor do I try to construct them any longer, for I know that my brain cannot get through in this way. But I rummage through the consequences as carefully as possible so as to learn about the range of applicability of this concept...".

Regarding Einstein's wish to detect appropriate fields of application, the Council offered him a rare opportunity to consult first-rate experimenters. His discussions with Warburg proved fruitful: they gave rise to a successful collaboration on the problem of photochemical decompositions[227] (Warburg undertook experiments which confirmed the validity of Einstein's law[228] of "photochemical equivalence").

Einstein was equally pleased by his discussions with Marie Curie, Langevin, Perrin and Wien. We will have much more to say about the impact of his contacts with the above French Council members (see section 3.1). However, it seems appropriate to point right now at a possible link between Einstein's contacts with Wien and his first nomination[229] for a Nobel Prize.

Indeed, it is known that Wien had a special interest in the ether-problem[230] and that he used every opportunity to discuss relativity theory. In particular, it has been reported that he discussed the matter with Rutherford in September 1910, on the occasion of the Congress of Radiology and Electricity[231], held in Brussels. Wien explained that Newton was wrong in the matter of relative motion. He then added *"But no Anglo-Saxon can understand relativity"*. *"No."* laughed Rutherford, *"they have too much sense"*.

Hence, it is more than likely that one year later Wien raised the ether-problem in his conversations with Einstein, and that, struck by the depth of the latter's reflections, he resolved to put his name forward for a joint Nobel Prize with Lorentz... Whether this really happened, or not, is still a question. But we have proof that Wien, who became the 1911 Nobel laureate for physics, proposed a few months later to award a joint Nobel Prize to Lorentz and Einstein for their work in the theory of relativity.

General impression and positive points

The following comment, made by Brillouin[232] during the final discussion, is indicative of the general climate at the end of the Council:

> "It now seems quite certain that it will be necessary to introduce into our physical and chemical concepts a discontinuity, an element varying by leaps, of which we had no idea a few years ago. How should it be introduced? This is what I see less well. Will it be in the first form proposed by Mr. Planck, despite the difficulties it raises, or in its second form? Will it be in Mr. Sommerfeld's form, or in some other form still to be found? I don't know... The uncertainty in which we find ourselves about the form and extent of the transformation that must be operated – evolution or complete overhaul – is a powerful stimulus. This concern will certainly haunt us for many weeks, and each of us will passionately focus on the solution of the difficulties that our discussions have shown to be inevitable and important in so many fields of physics and chemistry...".

Brillouin was right. Back in Paris, Poincaré set to work and succeeded, within a few weeks, in settling the question that had been at the centre of the debates: must Planck's hypothesis be considered as a necessary condition?

The Council had failed to resolve the dilemma. Aware of the necessity to save an essential tool of mathematical analysis, the differential equation, Poincaré (figure 18) had pleaded in Brussels for an alternative to the introduction of quanta. Reconsidering the question calmly, he changed his mind and showed[233] by a masterful demonstration that the quantum hypothesis couldn't be circumvented: energy quanta had to be accepted as a consequence of Planck's radiation law.

FIGURE 18: Henri Poincaré in 1911 (sitting, right hand side of the picture), see figure 1.

Eager to correct the position he had defended in Brussels, Poincaré sent his conclusion to the editors of the Council's report. It reached them in time to be added to the text in the form of a footnote, but the correction came too late to change Einstein's opinion. Convinced before anyone else of the necessity of the quantum hypothesis, the youngest member of the Council returned to Prague with the impression that Poincaré hadn't lived up to his reputation. These are his words in a message to Zangger[234] on November 15, 1911:

> "Poinkaré (sic) was simply negative in general, and, all his acumen notwithstanding, he showed little grasp of the situation. Planck stuck

stubbornly to some undoubtedly wrong preconceived opinions... The whole story would be a delight to diabolic Jesuit fathers...".

However, the importance of Poincaré's result did not escape Brillouin, who expressed his feeling in a letter to Lorentz[235]:

"Did you read Mr. Poincaré's penetrating note? Would Mr. Solvay have caused only this result, he would not have wasted his time...".

Indeed, Poincaré's authority was such that his conclusion imposed itself on Jeans, the British Council member who had opposed the idea of energy quanta, before becoming one of its most fervent defenders[236]...

Today, we can only agree with Brillouin: Poincaré's achievement of 1912 was the most convincing proof of the Council's success.

Other facts attest to the influence of the Solvay debates on the birth of atomic physics and on the emergence of quantum mechanics:

- Despite the failure of Planck's emission theory, his proposal contained elements that would reappear with the formulation of quantum mechanics (the need to accept the presence of elementary processes subject to a probability law; the existence of a zero-point energy for a quantum oscillator...).
- Following Sommerfeld's lecture, Marie Curie insisted on the presence of two distinct atomic regions: an internal region and a peripheral one. She based her conclusion on the result of scattering experiments involving alpha and beta rays. Rutherford, on the other hand, remained surprisingly silent about the alpha scattering experiments, carried out in his Manchester laboratory, which suggested the presence of an atomic nucleus. This prudence (in spite of the fact that his first conclusions were published[237] in May 1911) was apparently due to the need of additional evidence in favour of the nuclear hypothesis[238].
- The discussion of two Solvay reports, Planck's and Sommerfeld's, drew attention on an early atomic model based on Planck's quantum of action (the Haas[239] model).
- Niels Bohr, the founder of modern atomic theory, benefitted from the Council's discussions. He obtained first-hand information on the Solvay debates during his encounter with Rutherford, just after the Council (see Bohr's recollections[240] at the twelfth Solvay Council on Physics).
- Louis de Broglie's thesis of 1924, which extended Einstein's ideas about light to the world of matter, should be seen as a long-term consequence of the first Council in physics. Maurice's younger brother had been fascinated by the reading of the Council's provisional report. He immediately decided to become a physicist, and to strive to clarify the mystery of Planck's quanta[241]. His prediction of matter waves would be confirmed in 1927. The same year,

he would present his pilot-wave theory at the fifth Solvay Council on Physics, next to Schrödinger's report on wave mechanics and to the Born-Heisenberg report on quantum mechanics (see section 9.3).

To these particular consequences, we should add two elements of a more general scope:
- Before the Brussels meeting, physicists largely ignored the quantum problem (the term "quantum" didn't appear in papers, not even in Planck's). The Council brought the problem to the attention of the much larger community of physical chemists[242].
- The scoop of the Council – Kamerlingh Onnes' announcement of the phenomenon of superconductivity – marked the birth of a new, quantum related research domain: low temperature physics.

Press reactions

As noted by Eliane Gubin[243], Belgian newspapers didn't consider it useful to echo the holding of a *private* conference. The more so as it had been convened to discuss some obscure questions regarding thermal radiation and specific heats. In Brussels, there was no announcement of the event. Solvay, or people around him, may have preferred to remain discreet about this strictly confidential meeting.

Not so in Germany and in the Netherlands. Indeed, the *Berliner Tageblatt* published an announcement of the Council on October 27, 1911, presumably at Nernst's instigation. One indication of his involvement was the emphasis on Solvay's role as initiator of the conference. Another was the fact that the paper contained the names of van der Waals and of Goldschmidt (Nernst, as we know, had put pressure on van der Waals, hoping to convince him to attend the meeting). Goldschmidt's name may have been cited to deny the embarrassing absence of Belgian Council members. Last but not least was a statement in the Tageblatt which corresponded in every way to Nernst's state of mind:

"Future generations will learn from the conference report how the brightest minds of the time attempted to meet the challenges of mechanics and atomism."

The announcement of the *Berliner Tageblatt* was taken up in the Netherlands, where it appeared in the *Algemeen Handelsblad*.

Nothing of the kind happened in Belgium, where the newspapers chose to ignore the Council, with the exception of the "Curie-Langevin affair", which caused rumours and unwanted publicity, but seemed likely to attract the attention of a large audience.

The scandal broke out in Paris on November 4, 1911, the day after the end of the meeting, with the publication by the *Journal de Paris* of a report with the following introductory lines:

> *A love story*
>
> *Madame Curie and Professor Langevin*
> *The fires of Radium which radiate so mysteriously on everything prepared us a surprise: they started a conflagration in the hearts of the scholars who study their action with tenacity, while the scholar's wife and children are in tears...*

The affair was relayed to Germany on November 5, 1911, by the *Berliner Tageblatt*, first as a telegram, then in more detail the following day.

As a contrast, Lindemann's note "*A most curious Congress*" only appeared in the *Tageblatt* in February 1912. In this note Lindemann stressed the importance of the conference, listed the introductory reports, and quoted the names of the Council members, with the exception of Marie Curie's... However, she was cited elsewhere in the paper, as the only invited woman.

In Paris, the Curie-Langevin affair continued to cause a stir[244]. On November 5, it was the turn of the *Petit Journal* to headline:

"*A novel in a laboratory: the affair of Mrs. Curie and Mr. Langevin.*"

FIGURE 19: Paul Langevin in 1911, see figure 1.

Journalists rushed to the Collège de France and to Curie's laboratory to interview Langevin (figure 19) and Marie Curie. Having been told that the two "had gone to Brussels to attend a Congress", they alerted their Belgian correspondents.

Two days later, *Reuter* announced that Marie Curie was to receive the 1911 Chemistry Nobel Prize for her discovery of radium. After sharing the 1903 Physics Prize with Henri Becquerel and Pierre Curie for the discovery of radioactivity, Marie had the extraordinary privilege to become history's first scientist to receive two Nobel Prizes. But this supreme distinction was eclipsed by the other, much more sensational news. The "affair" was in everybody's mind, and the scandal was growing.

On November 9, Perrin and the mathematician Emile Borel, friends of the Curies, were summoned by the police prefect. Meanwhile, Marie and her daughters, had been forced to flee the demonstrations in front of their home in Sceaux, and had found refuge in Paris with the Borel family.

On November 23, a paper entitled "*The truth about the Curie-Langevin scandal*" appeared in *l'Oeuvre*, a xenophobic and anti-Semitic journal directed by Gustave Téry, a former classmate of Langevin. It reported the summons issued at Langevin by his wife's lawyer, where some of Marie's letters were cited...

The Perrin and Borel families tried to rehabilitate their illustrious friend among their professional relations. Langevin, on his part, wasn't defeated. He challenged Téry to a duel, and went on November 25 to the Parc des Princes with two witnesses: Paul Painlevé, a well-known mathematician, and Albin Haller, the director in Paris of the ESPCI, and co-founder of the IACS. But the duel was cut short: Langevin and Téry decided to lower their pistols...

In Stockholm, the scandal provoked concern, especially with Arrhenius who had made efforts to promote Marie's candidacy. On December 1, he asked her to deny the charges, and advised her to abandon the idea of going to the Nobel Awards ceremony. Fortunately, Arrhenius' advice remained without effect. Reassured by a confidential message[245] from his rival, Mittag-Leffler, Marie decided to ignore the warning and went to Stockholm to receive her Prize.

In Belgium, liberal newspapers took sides with Marie Curie, declaring her victim of xenophobia and anti-feminism[246]. *L'Étoile belge*, a daily paper engaged in a local anti-clerical struggle, took her defence and published this ironic statement:

> "*Madame Curie is wrong to be both learned and famous, to want to enter the Institute and to outshine a few equally ambitious men... Today, a scandal-paper seizes her name, her private life, and reveals – appalling detail – that the ideal companion and student of the great Curie, is perhaps only a woman like many others, subject to deficiencies... Here is Mme Curie, dirtied at will, and the members of a large coterie are rubbing their hands: the unbeliever is forever excluded from the Institute.*"

Newspapers in France echoed the turmoil in the academic world. However, none of these journals had access to the reactions of the scholars who had been close to Marie Curie and to Paul Langevin during the Council. Here is a confidential message sent by Rutherford[247] to his Austrian colleague Stefan Meyer:

> "*No doubt you have learned that Mme Curie is far from well and will require a long rest before she will be able to get to work again... I trust that the rumours that you have heard of the cause of Mme Curie's illness are unfounded. I certainly sympathize very strongly with her in the misfortunes that have come to her in recent years...*".

And here is Einstein's reaction in a letter[248] to his friend Heinrich Zangger, professor of toxicology at the University of Zurich:

> "*I returned last night from Brussels where I spent much time with Perrin, Langevin and Marie Curie, and became quite enchanted by these people.*

The latter even promised to come visit us with her daughters. The horror story that was peddled in the newspapers is nonsense. It has been known for quite some time that Langevin wants to get divorced. If he loves Mme Curie and she loves him, they do not have to run off, because they have plenty of opportunities to meet in Paris. But I did not at all get the impression that something exists between the two of them; rather, I found all three of them bound by a pleasant and innocent relationship. Also, I do not believe that Mme Curie is power-hungry or hungry for whatever. She is an unpretentious, honest person with more than her fill of responsibilities and burdens. She has a sparkling intelligence, but despite her passionate nature, she is not attractive enough to represent a danger to anyone...".

Yet, Einstein was soon forced to face the facts. Deeply shocked by the xenophobic character of the attacks on Marie, he sent her this message of support[249]:

Don't laugh at me if I write to you without having anything substantial to tell. But I am so enraged by the baseness with which the public has dared to take an interest in you these days, that I absolutely must share this feeling with you.

Yet, I am convinced that you uniformly despise this scum, whether it lavishly praises you, or satisfies its insatiable appetite for sensationalism.

I want to tell you how much I admire your intelligence, your energy and your integrity, and how happy I am to have been able to meet you in Brussels.

Anyone who doesn't belong to this band of reptiles can only rejoice, today as in the past, to feel close to people such as you and Langevin, sincere and true characters with whom every contact is perceived as a privilege.

Do not read the garbage in case the scoundrel continues to take care of you, rather abandon it to the reptiles for which it was made.

My fondest memory to you, Langevin and Perrin, your very affectionate,

Albert Einstein

One man found himself in a cruel embarrassment: the Council's tactful Chairman Lorentz, a declared partisan of the emancipation of women. It was on his advice that Marie Curie had been included in the "Select Committee" which had met on the last day of the conference... Hence, it was up to him to ensure that the scandal wouldn't reflect on Solvay.

But what should he do? Was it desirable to keep Marie in the Institute's "Scientific Committee"? Before making a decision, Lorentz decided to get advice from the other French Committee member, Brillouin.

These are Brillouin's words in a letter[250] to Lorentz of January 29, 1912:

Dear Sir,

Your letter touches on one of the most distressing questions that have faced us for a long time. I would have preferred that this lamentable adventure remained ignored; but you had indeed a duty to inform yourself, and since you have read the documents that have been published, the main one being the unfortunately authentic letter published by "l'Oeuvre", I cannot dispense with telling you what I think, and why Mr. Solvay deserves that your decision be taken knowing the moral disturbance which is currently shaking our academic world (...).

I knew Langevin as a student at the École Normale, and I immediately thought of him what I still think today: that he was destined to become the main physicist in France and that he had to be directed to the Collège de France, so that my stepfather, Mr. Mascart, would take him as his deputy.

We have known, my wife and I, the young household Langevin during the first years of marriage; Langevin showed affectionate dedication. Mrs. Langevin was a mother and a wife to whom no criticism could be made. Of modest origin, like Langevin himself, she courageously supported with him the heavy bilateral family responsibilities. My wife, who saw her informally at the time – we lived from Easter to autumn in the same village near Paris to offer fresh air to our children – never heard her say a word that a well-educated woman wouldn't utter, nor express an indelicate feeling.

I am now told that from that time on, there were violent arguments between them, with rude words on the part of Mrs. Langevin, and that Langevin was very unhappy... It is quite possible, but he didn't look like it; he didn't blush because of his wife. Both were warmly welcomed everywhere, and four children, the last of which is about three years old, seem to prove that there was nothing irremediable in their disagreements, unknown to all of us at the time.

This brings us to the deplorable accident that cost Pierre Curie's life. All of Curie's personal friends put their dedication at the service of the unfortunate widow. She dreaded the ordeal of public teaching. No one was surprised that she resorted to Langevin's advice. By an unfortunate coincidence, from this time dates – apparently at least – Langevin's disaffection for his wife (...).

What to do? I am quite unable to tell you anything about it. We must believe that it is not an easy matter, because the men who are at the head of the Paris University are – so it seems – cruelly embarrassed. Their relations with Mr. R. Poincaré, lawyer of Langevin and currently President of the Council of Ministers, certainly enable them to know the bottom of things, so as to avoid the public scandal of a trial, with letters, witnesses, etc...

What did they decide? I don't know. I systematically avoided any meeting that would have forced me to publicly express my feelings. I contented myself with making them known, and quite vividly, to Perrin. I could

> *not avoid the assemblies of our College; but Langevin understood and we easily manage not to meet, although we both attend. The Collège de France is by definition a collection of independent scholars. We can ignore each other without the College suffering, but it is one of the biggest sorrows of my life (...).*
>
> *You are, Sir, by your character and your nationality, the only foreigner in whose esteem, sympathy and righteousness, I trust enough to express my feelings on this painful subject, without fear of being badly judged. I take responsibility for what I write, but it would be painful to me, as a Frenchman, if my feelings were given international publicity. I therefore want this letter to remain confidential...*

We will see that despite Brillouin's reservations, Lorentz kept Marie Curie as a member of the Institute's Scientific Committee. His decision was facilitated by the reassuring news from Paris: the rumours of scandal died down before the end of 1911.

However, Marie's health had been undermined by the violence of the attacks. On December 29, before settling in the apartment she had rented in Paris, she was rushed to the hospital... Fortunately, she soon learned from the terms of the separation judgment of the spouses Langevin that the scandal had been defused, thanks to the skilful intervention of Langevin's lawyer[251], Raymond Poincaré (a cousin of Henri). The text of the judgment didn't mention Marie Curie's name.

Solvay's attitude in the aftermath of the Brussels meeting

Despite the disappointment of some members, the success of the Council is measured today by the magnitude of its consequences. Its impact would soon be reinforced by the action of Solvay's international institutes: ISIP and ISIC.

We will report on the birth of these institutes in the second part of our story.

However, it seems appropriate to end this first part with a discussion of the Council's impact on the man who summoned it.

We have seen that Solvay realized at the end of the meeting that his guests had failed to resolve the crisis. After a week of discussions, these eminent scientists only agreed on one point: the classical theories' inability to account for the facts.

This meagre result could have demotivated the conference's host. But it wasn't the case. Far from being discouraged, Solvay felt comforted in his faith that the crisis could only be resolved by means of a totally different approach, in particular his own...

In other words, the Brussels discussions were a stimulating factor. They encouraged Solvay to pursue his research with redoubled ardour. More eager than ever to complete the work described in the "Preliminary Study" he had distributed to his guests, against the advice of Tassel, he decided to call on the services of younger collaborators on a permanent basis.

In this regard, it is worth to recall a fact that hasn't been highlighted so far: the absence at the Council of Solvay's faithful advisor for physics, despite his former involvement in the preparations for the meeting (see section 2.5 and Herzen's message of June 3, 1911). Tassel's absence is all the more surprising as he had been the co-author of Solvay's *Gravitique*[252] written in 1887, which served as a basis for the drafting of the "Preliminary Study".

It raises an obvious question: What was the reason of Tassel's withdrawal from the Council and his replacement by two newcomers: Herzen and Hostelet?

An answer can be found in Tassel's anonymous *"Notes on the work carried out by Ernest Solvay from 1857 to 1914"*, printed in 1920, a copy of which is kept at Teyler's Museum in Haarlem.

Here are Tassels's relevant words (pp. 65-66):

"Two employees of the Solvay Company Research Laboratory, Messrs. E. Herzen and G. Hostelet, who since 1909 were involved in the study of certain problems related to Mr. Solvay's personal researches, theoretical or applied, were called to collaborate with him, first part-time, then full-time, in the framework of his general theories. Until then, Mr. Solvay had mainly been assisted, in the various directions of his activity, by a collaborator whose support had become insufficient to ensure, at short notice, the completion of his studies and the scheduled publications.

On the other hand, Mr. Solvay needed a more flexible collaboration, less subject to critical and classical thinking, i.e., more open to his personal views, that only younger people could provide...".

Tassel's letters to Solvay, written during WWI, contain elements which confirm his wish to be no longer associated with the industrialist's increasing desire to burn all the steps of a normal research. We may therefore assume that the retired professor chose to step aside in anticipation of the Council, and to give way to two *"younger and more receptive"* physicists.

As to Solvay, it seems clear that by freeing himself from Tassel's criticism, he felt confident that it would liberate his research from unproductive conservatism. Furthermore, an appointment of younger collaborators would allow him to give free rein to his boldest ideas. Solvay's appointment of the two physicists who assisted him since 1910 intervened at the end of 1911. Soon after the end of the Council, he signed an exclusivity contract with Herzen and Hostelet, offering them significant financial advantages[253]. Six months later, he confirmed their position by including a clause in ISIP's statutes, stating that *"Messrs. Herzen and Hostelet who attended the 1911 Council"* would be invited to attend future scientific meetings, as Solvay's official representatives (see section 4.2).

On the other hand, we will see that Solvay's old and faithful advisor for physics would continue to enjoy the industrialist's fullest confidence for actions other than those connected with his personal research. We will see in section 4.4 that Solvay called on Tassel's services in 1912, asking him to join Héger, in order to finalize the foundation of the International Institute of Physics. We will also see (sections 4.3 and 4.4) that a few months later Tassel became Solvay's representative, in charge to find an agreement with the leaders of the IACS for the creation of International Institute of Chemistry.

Regarding Solvay's personal research, we will see that Tassel's withdrawal had notable repercussions on the industrialist's program and on his general behaviour. Deprived of the moderating influence of his former adviser, Solvay embarked on increasingly hazardous paths that took him further and further away from ordinary, well-established scientific practices.

We will report on this unreasonable aspect of Solvay's personality in the second part of this book. However, it is worth underlining right now the striking contrast between the two images that emerge from the initiatives of this extraordinary man.

On the one hand, we have the image of a huge success, due to an unquestionable insight and greatness of mind. Thanks to the freedom enjoyed by his international foundations, ISIP and ISIC, Solvay's passion for science had a strong and lasting impact on the development of modern physics and chemistry.

On the other hand, we have the image of an isolated investigator, carried away by an unrealistic dream, who gets lost in endless speculations, without any real chance of success.

Faced with this second, distressing aspect of Solvay's activity, there is a question which comes to mind: Why did this prudent, experienced man, who rubbed shoulders with the most famous scientists of his time, persist in his choice to confine himself in an obstinate, almost solitary research, without taking expert advice?

Solvay's refusal to discuss his theory with the Council members was noticed, as we have seen, by Sommerfeld, who reported this fact in a letter to his wife. We will also see (section 9.4) that this strange attitude astonished Lorentz, the tactful theorist who would have been most able to enlighten him.

One aspect of Solvay's personality will become clear in the next chapters: his fear of submitting his ideas to theorists, while agreeing to share them with experimenters.

A possible explanation of this curious attitude may be found in the industrialist's early contact with Belgium's most skilful and respected chemist: Jean Servais Stas.

We mentioned the fact that Solvay consulted Stas about a possible equivalence between mass and energy, and that he didn't get any support. Much more damaging was Stas' reaction to Solvay's idea of producing soda from

An unprecedented project

salt and ammonia on an industrial scale. Responding to the latter's questions regarding the feasibility of his plan, the author of the first table of atomic weights had been categorical[254]: *"the Solvay process is too delicate to allow industrial application"*.

It is hard to believe that after having brought a scathing denial of Stas' affirmations, the successful producer of ammonia-based soda wouldn't have sworn to give no credit to the opinions of celebrated experts... Wasn't it this early experience which forged Solvay's mistrust of first-class specialists, who with the best intentions, were likely to demotivate the heralds and architects of future progress?

UNEXPECTED CONSEQUENCES OF THE COUNCIL

We saw that the first Council in Physics – we will henceforth call it "Solvay I", resulted from very different motivations, which however all agreed, thanks to an impressive combination of circumstances.

Walther Nernst, father of the third law of thermodynamics, should forever be remembered for having thought of a "Scientific Council" and for having suggested an unprecedented working method[255]. This method, unduly called "Solvay method", would become the trademark of the famous Councils in physics and in chemistry.

Ernest Solvay, the self-taught visionary with a passion for science, which he practiced in a very personal way, deserves an equal credit for having understood that physics was in crisis, and for having given substance to Nernst's idea.

We will now see that "Solvay I" had various consequences. Some of them are well-known, in particular Solvay's foundations in physics and in chemistry. Others have never been the subject of a detailed report. They will therefore be the starting point of the second part of our story.

Chapter 3
A game of musical chairs

The Brussels meeting had an immediate impact on the professional career of two prominent physicists: Lorentz and Einstein. The Council can therefore be described as the start of an impressive professional merry-go-round.

3.1 Impact on Einstein's career: from Prague to Zurich

Einstein left Brussels with the feeling that he hadn't convinced the Council members of the need to carry out an in-depth reform of Maxwell's electromagnetic theory. He knew that his colleagues weren't ready to accept a point that he saw as being firmly established: the fact that in each periodic process energy could only be exchanged by quanta[256].

But Einstein was delighted to have met French and British physicists. Until then, his professional meetings had mostly been limited to the German scientific sphere. One exception had been an invitation to Geneva in 1909 to receive an honorary doctorate; another had been an invitation to Leiden in February 1911. The links he had been able to forge at the Council with Marie Curie and with Paul Langevin would give birth to a true friendship.

On November 15, 1911, Einstein sent this enthusiastic message[257] to Heinrich Zangger:

> "It was most interesting in Brussels. In addition to the French Curie, Langevin, Perrin, Brillouin, Poinkaré (sic) and the Germans Nernst, Rubens, Warburg, Sommerfeld, also Rutherford and Jeans were there. And of course, also H. A. Lorentz and Kamerlingh Onnes. H. A. Lorentz is a marvel of intelligence and tact. He is a living work of art. In my opinion he was the most intelligent among the theoreticians present...".

Although concerned about professional matters when taking the train to Brussels, the youngest member of the Council was far from imagining the impact that his presence at the meeting would have on the development of his academic career.

The Utrecht chair

Let us go back to the end of August 1911, when Einstein considered leaving the German University of Prague where he held a chair since April 1911. The idea came to him on August 20, when he learned[258] from his Dutch correspondent Willem Julius that a chair was vacant at the University of Utrecht. Julius wanted to know if an offer from the Faculty would have any chance to attract him to Holland.

Einstein's first reaction was to ignore the news. The prospect of leaving Prague after such a short time placed him in a complicated situation. However, the idea of getting close to people such as Julius, Lorentz and Kamerlingh Onnes soon gained ground. Little by little, it strengthened his desire to leave Prague...

Einstein hadn't forgotten that the Eidgenössische Technische Hochschule (ETH) in Zurich, where he had been trained, had refused him a post of assistant at the end of his studies. He now realized that the Utrecht offer could increase his chances of an appointment in his former school, providing him with a revenge he had longed for. But the case was delicate: any decision regarding the ETH had to be taken in Bern. Einstein decided to discuss the matter[259] with his friend Zangger who was passing through Prague. Zangger had close connections with Ludwig Forrer, a federal Councillor. He agreed to conduct negotiations[260] with the federal authorities in Bern.

Einstein's additional problem was to adopt an adequate attitude towards the people in Utrecht. Indeed, Julius returned to the charge on September 17, this time on behalf of the Faculty. Pressed to give an answer, Einstein sent a letter to Zangger, asking him what his chances were in Zurich[261]... Five days later, he told Julius[262] that he was thinking of coming to Utrecht, but that he had promised, before leaving for Prague, to return to Zurich in the event of an offer from the ETH.

Soon afterwards (from September 24 to 30) Einstein took part in the "Naturforscher Versammlung" in Karlsruhe, where he met again Zangger, this time in the company of Pierre Weiss, the director of the Physics Institute at the ETH.

On October 11, Julius returned to the news. Seeking to gain time, Einstein asked him for a period of reflection, explaining that he no longer believed in an offer from the ETH... However, having learned a few days later that he could count on support in Zurich, he urged Zangger to be diligent[263], as the Dutch were getting impatient.

A game of musical chairs

Clearly, for Einstein the second half of October 1911 was a difficult time. Increasingly irritated by the delays in Bern and in Zurich, he regretted his commitment to Nernst and to the "*Konzil*", which forced him to interrupt his work. On October 21, he felt the need to confide to his friend Michele Besso[264]:

> "*I really did not have a single free moment. I was away 3 weeks, first in Karlsruhe at the Naturforschervers and then in Zurich, where I had to give 8 lectures at the vacation course. Add to this the many shop-talks and personal obligations. But now – once the witches' sabbath in Brussels will also be over – I will be my own master again, except for my courses...*".

Unfortunately, things dragged on. The appointment at the ETH was still pending when Einstein arrived in Brussels. He spoke to Lorentz, who had met Julius during summer, and more recently before the meeting. The chair in Utrecht was mentioned in their conversations[265]. The Council's chairman thought he might convince Einstein to come to Holland. But the bottom of the question wasn't addressed. Lorentz, who didn't know the details of Einstein's negotiations with Utrecht, preferred to remain discrete. Einstein, who wasn't informed on Lorentz's position decided not to come forward. This lack of clarity on both sides was a source of misunderstanding[266].

Just after the Council, Einstein went to Utrecht. He told Julius that he would give him a final answer after his return to Prague, where a letter from Zurich should be waiting for him[267]... Back home, he noticed that it wasn't the case: no letter from Zurich. Overwhelmed by the situation, Einstein decided not to go to Holland. On November 15, he communicated his decision to Julius[268], explaining that he had written to Debije, a Dutch physicist who occupied his former position in Zurich, and would be glad about an appointment in Utrecht. The same day, he sent a letter to Zangger[269], asking him not to deal any longer with his appointment in Zurich.

The next day he received a letter, apparently from his friend Marcel Grossmann, with the information that the ETH wished to appoint him as professor in theoretical physics. The fact was soon confirmed by Weiss: Zangger's efforts had paid off.

Einstein hastened to announce the news to Julius, asking him not to publicize Debije's appointment in Utrecht[270] (it was essential to keep the pressure on the authorities in Bern, who had decided in his favour only out of fear of seeing him settle in Utrecht).

Thus, towards the end of November 1911, Einstein's return to Zurich was no longer in doubt. However, the news of his official ETH nomination[271] on January 30, 1912, by the federal authorities would only reach him in February 1912.

Influence of the Council

On the last day of the Brussels meeting, Einstein didn't have the slightest idea of the support that the "witches' sabbath" was going to bring to his candidacy in Zurich. He couldn't imagine that one of his biographers[272] would write one day: "*Einstein's attendance at the Solvay Congress of 1911 had repercussions which decisively affected the rest of his life*".

A first repercussion materialized a few days after the end of the Council. Following Zangger's conversations with Weiss, the latter contacted two of his French colleagues who had taken part in the Council[273]: Marie Curie and Henri Poincaré. Both had been impressed by Einstein's performance in Brussels. They highly recommended him for the position in Zurich.

These are Marie Curie's words in her letter[274] to Weiss of November 17, 1911:

> "*I much admire the work which Mr. Einstein has published on matters concerning modern theoretical physics. I think, moreover, that mathematical physicists are at one in considering his work as being in the first rank. At Brussels, where I took part in a scientific conference attended by Mr. Einstein, I was able to appreciate the clearness of his mind, the sharpness with which he interprets the facts, and the depth of his knowledge. If one takes into consideration the fact that Mr. Einstein is still very young, one is justified in basing great hopes on him, and in seeing in him one of the leading theoreticians of the future. I think that a scientific institution which gave Mr. Einstein the means of work which he wants, by appointing him to a chair in the conditions he merits, could only be greatly honoured by such a decision, and would certainly render a great service to science.*"

And here are Poincaré's comments[275]:

> "*Mr. Einstein is one of the most original minds I have known. Despite his youth, he has already taken a very honourable rank among the first scholars of his time. What we should especially admire in him is the ease with which he adapts to new concepts, and knows how to draw all the consequences. He doesn't remain attached to classical principles, and, in the presence of a problem of physics, is quick to consider all the possibilities. This is immediately reflected in his mind by the forecast of new phenomena, capable of one day being verified by experiment. I do not mean to say that all these forecasts will withstand the confrontation with experiment the day this confrontation becomes possible. As he searches in all directions, we should on the contrary expect that most of the paths he takes will be dead ends; but we must at the same time hope that one of the directions he has indicated is the right one, and that's enough. This is the way to proceed. The role of mathematical physics is to ask the right questions; it is only experiment that can answer them. The future will*

show more and more the value of Mr. Einstein, and the university which will have had the wisdom to appoint this young master is sure to derive much honour from him."

Marie Curie and Poincaré weren't the only members to appreciate Einstein's talents. We already mentioned Wien. Most striking are the attempts that were made in various countries to attract the young prodigy of German science[276]:
- In Holland, Julius' efforts were followed by Lorentz's attempt to welcome Einstein as his successor in Leiden, a project which started in February 1912 (and on which we will have more to say in section 3.2).
- In June 1912, Warburg, president of the PTR, made an offer for an appointment at his Institute in Charlottenburg.
- One month later, it was Hasenöhrl who tried to attract Einstein to Vienna, an effort that would be eclipsed by the vigorous actions undertaken in Berlin (and to which we will come back in section 3.3).

3.2 The imbroglio of Lorentz's succession

At the time of the Council's opening, Einstein wasn't the only member concerned about his professional future. Lorentz, holder for more than thirty years of a chair at the University of Leiden, nurtured the hope to be able to devote himself more to research, both theoretical and experimental (a wish constantly postponed because of his heavy teaching duties).

A decisive event had taken place in 1909, when Lorentz was approached by Teyler's Foundation in Haarlem with a proposal to appoint him as the *curator* of its physics laboratory. The proposal implied the eventual renouncement to his Leiden chair in theoretical physics, and his installation in Haarlem. It was a huge step, but the idea appealed to Lorentz. The new job would leave him a lot of freedom.

The project materialized in January 1910, when Lorentz took charge of the Teyler laboratory, in agreement with the Faculty, the University in Leiden and the Ministry.

It had been agreed that he would keep his position in Leiden for a period of time, about four years[277], and that his ordinary professorship would be transformed, upon departure, into an extraordinary one. A new holder of the chair being appointed, he would only have to give one course on a subject of his choice.

The arrangement proved satisfactory. As soon as he took up his post, Lorentz set to work reorganizing the Teyler laboratory. A physicist had been hired to assist him in his experiments. The new curator didn't imagine that an unexpected obstacle would arise a few months later, preventing him from carrying out his research program.

The position in Utrecht

In anticipation of his change of status (from ordinary to extraordinary professor), Lorentz felt compelled from the summer of 1911 to take care of his succession.

The case presented itself in a favourable light. Competition wasn't to be feared.

No theoretical chair in physics was vacant in the Netherlands... However, the situation changed on August 8, with the death at Utrecht of Professor Cornelis Wind. Julius, who was in charge of finding a successor, thought immediately of Einstein and sent him a confidential message on August 20. He also met Lorentz to discuss the vacancy in Utrecht.

Shortly afterwards, Lorentz received a visit from Debije. Favourably impressed by Debije's personality, he communicated his impression to Julius. Debije, who had been in contact with Einstein, told Lorentz that the latter had gone to Prague because of the high salary and that people in Switzerland were trying to lure him back to Zurich with the prospect of an appointment at the ETH.

Lorentz's first encounter with Einstein dated back to February 1911. The young physicist had come to Leiden upon invitation from students. He and his wife had been invited to stay with Lorentz, something that wasn't in Lorentz's habits[278].

Lorentz brought his guests to Haarlem to see the Teyler Museum and the foundation's physics laboratory. He probably told them about his intention to settle there three years later[279], unaware of the fact that this schedule would be upset by the holding of the Council.

When he arrived in Brussels on October 29, Lorentz spoke with Einstein. He had been informed by Julius of his attempt to attract Einstein to Utrecht (a fact confirmed by Julius just before the Solvay meeting). The vacancy of Wind's chair was discussed, but Lorentz's words were unclear: Einstein mistakenly believed that Lorentz favoured a Dutch candidate[280], a preference which should work in favour of Debije... Held by his commitment to Julius, Lorentz refrained from mentioning the future vacancy of his own chair, and from expressing the hope of welcoming Einstein one day in Leiden as his successor.

Back to Leiden, the Council's chairman took steps to speed up the opening of his succession. On November 21, he sent a letter to the Faculty, asking a revision of the 1912 budget in view of a reduction of his mandate from ordinary to extraordinary professor.

Lorentz's change of mind – he now wished to retire at least one and a half year earlier than expected – was a direct consequence of the mission that Solvay had entrusted to him a few weeks earlier: the design and future direction of an International Institute for Physics.

In an attempt to justify the haste with which he submitted his request to the Faculty, Lorentz invoked the extra work caused by his presidency of the physics section of the Amsterdam Academy (a position held since 1910). He nevertheless revealed his true motives in la letter of November 21, 1911, to J. D. van der Waals[281], curator of the University and an invitee to the Council:

> "To you, personally, I may say that one of the elements which contributed to the increase in my workload is a consequence of the Brussels congress. Mr. Solvay wants to promote progress in physics and in chemistry by means of a new foundation (principally through the financial support of laboratories). By virtue of my position as president of the Council, I have been asked to lead the discussions with him and with other members. I predict it will cost me a lot of effort.
>
> It is of course a secret, but that you, as a Council member, are authorized to know."

Two days later, Lorentz received unrelated news from Einstein[282], who expressed his embarrassment of having to decline the Utrecht chair:

> "I write this letter with a heavy heart, like a person who has done some kind of injustice to his father The decision I had to make was very difficult for me. Now I admit to you quite frankly that one of my main worries was that I did not know whether you would think it proper if a foreigner were to come to Utrecht. After all, I could not ask you directly, and no one else would have been able to tell me. Add to this my conviction that the young Debije, who is Dutch, after all, is at least my equal as far as talent is concerned.
>
> You have probably sensed that I revere you beyond measure. If I had known that you wanted me to come to Utrecht, I would have gone there. But it is easy to understand why I did not dare to ask you. So there is only one thing left to me: I beg you earnestly not to be angry with me for the way I acted. It is a punishment enough that I am now denied the opportunity of meeting with you more frequently. But if you want to remain on friendly terms with me in spite of what happened, I shall come to Holland from time to time to talk with you. Also, I would like to ask you to give me the great pleasure of having you as my guests if you come to Switzerland with your family...".

With this emotional message, it became clear that Lorentz's preferred candidate wouldn't come to Holland. Indeed, Einstein's last words were unmistakable proof that he had opted for Switzerland[283].

In his reply of December 6, 1911, Lorentz told Einstein that he didn't hold it against him for the Utrecht affair, and that he consoled himself with the idea that *"he would also accomplish great things in Zurich"*. However, he carefully refrained from making any allusion to his own succession (Lorentz couldn't prejudge the decision of the authorities in Leiden, nor that of the Minister).

The next day, Lorentz was informed that a ministerial proposal, in accordance with his request, had been approved by the Faculty. He immediately sent a message to Einstein[284], asking him if he was *"absolutely sure"* to go to Zurich; again without giving reasons for his insistence, but with the indication that he expected a quick response (the Minister's final decision hadn't yet been communicated).

Einstein, who wasn't aware of Lorentz's intentions, associated the message with the position in Utrecht, a case that didn't concern him anymore. Forced to be totally clear, he told Lorentz[285] that he had accepted an offer from the ETH, and that he couldn't backtrack out of consideration for a friend from Zurich who had gone at great lengths to support his candidacy.

The Minister's decision fell on January 5, 1912. It went in the desired direction. One month later, Lorentz's succession was discussed in a Faculty meeting. This is part of the meeting's report[286]:

"Mr. Lorentz observes that special circumstances call for diligence. He says that there is a person, Professor A. Einstein from Prague, whose merits as a theoretical physicist are so exceptional that he feels obliged to propose to the Faculty to give him the mandate of ordinary professor. He adds that no Dutchman will have reason to feel hurt by this choice...".

On February 13, the day after the meeting, Lorentz finally found himself authorized to make an offer to Einstein on behalf of the Faculty. He therefore hastened to send him the official news, together with this personal message[287]:

"As for me personally, I cannot tell you how enticing the prospect of maintaining a constant contact with you at work would be for me. If it were granted to me to welcome you here as my successor and as my colleague at one and the same time, it would be the fulfilment of a wish that I have long cherished in private but have not been free to voice before...".

Einstein reacted on February 18, explaining that Lorentz's letter had put him in a *state of turmoil*, but that he couldn't leave the path he had entered into by signing a contract with the ETH. Echoing Lorentz's emotional message, he declared[288]:

"And now the most admired and the dearest man of our times offers me a place close to him, in that he holds out to me the prospect of a friendly personal relationship.

I can think of nothing more beautiful than to experience the problems and developments in our mysterious science in conversation with you. My feeling of intellectual inferiority with regard to you cannot spoil the great delight of such conversations, especially because the fatherly kindness you show to all people does not allow any feeling of despondency to arise.

However, to occupy your chair would be something inexpressively oppressive for me. I cannot analyse this in greater detail, but I always felt sorry for our colleague Hasenöhrl for having to occupy Boltzmann's chair...".

For Lorentz, things were now perfectly clear: he had to look for another successor. Unfortunately, he couldn't call on Debije, who had been offered the position in Utrecht (Lorentz wouldn't dream of cutting the grass under his colleagues' feet).

However, unexpected news would soon change the course of events.

The Lebedew affair

As Lorentz tried to find a successor, he received an alarming letter from Russia. It was sent to him by the physicist Piotr Nikolaevich Lebedew[289], who informed him of the dramatic situation of his laboratory. About hundred professors and lecturers from Moscow University had been forced to resign after a brutal intervention of the police. Lebedew had found himself in need of appealing to the generosity of Muscovite businessmen. Thanks to their help, he had been able to set up his research team in private premises. Having learned that his Dutch colleague had been asked to direct the scientific committee of a newly created "International Institute for physics", he had felt it his duty to send him a request for support.

Lorentz was extremely surprised by Lebedew's request[290], because the Institute hadn't yet been founded... No decision had been taken so far about the granting of subsidies.

A conclusion was obvious: there had been a leak from the Select Committee, which had met with Solvay on November 3, 1911. It could have reached Lebedew directly from a Committee member or *via* an informed participant to the Council. It remained to identify Lebedew's informant.

Lorentz suspected Sommerfeld. He sent him a letter (which has unfortunately been lost) and received a reply[291] which contained the address of Paul Ehrenfest (figure 20), a physicist born in Vienna and working in St. Petersburg.

Ehrenfest had been staying a short time in Munich, and was due to return to Russia in mid-March 1912. Sommerfeld admitted that he had spoken to Ehrenfest of the planned "International Institute for Physics". He took advantage of his letter to Lorentz to praise Ehrenfest, describing him as *"even more fascinating in his conversations than in his writings*[292]*".*

One question should still be answered: How did Lorentz realize that Sommerfeld was responsible for the leak?

FIGURE 20: Paul Ehrenfest in 1921 (standing, right hand side of the picture), see figure 41.

Everything has to do with Ehrenfest, who worked with his wife, Tatiana Afanassieva, but had no position in Russia. Looking for a job, the young Austrian had embarked on a European tour[293]. In Prague, he met Einstein, who would become his lifelong friend. In Berlin, he visited Planck, who told him about the Council. Ehrenfest was informed that his work on the quantum hypothesis[294], published shortly before the meeting, had gone completely unnoticed... Worse still, he learned of an article, published after the Council, in which Poincaré had come to the same conclusion as the one reached in his paper.

Shocked by the news, Ehrenfest sent reprints of his work to Poincaré and to the other Council members. Lorentz received the document in the beginning of February 1912. He noticed that it had been posted in Munich, and therefore suspected a link between Ehrenfest's mailing and Lebedew's call for help... It all pointed to the only Munich member of the Council.

So, it was Sommerfeld who got informed of the discussions that had taken place with Solvay on November 3, 1911. He was the source of the rumours of subsidies that had spread from Munich to Moscow through Ehrenfest[295].

A providential successor

The "Lebedew mystery" having been cleared up, Lorentz decided to inquire about the Russian physicist's situation. It sufficed to ask Ehrenfest, whom he had known as a student in 1903, and who was still touring in Europe. But Lorentz needed to know more about the progress in Solvay's foundation plans. He therefore decided to wait for Ehrenfest's return to Russia. He would send him a letter using his St. Petersburg address (communicated by Sommerfeld).

Thus, after waiting several weeks, Lorentz wrote to Ehrenfest[296] on April 20, 1912. His decision was prompted by two developments:
- He had received the day before a letter from Solvay[297], urging him to obtain the adhesion of the Select Committee to the Institute's draft statutes. The industrialist clearly wished to speed up ISIP's foundation process.
- He had recently been informed of Lebedew's death.

Confronted with the disappearance of the famous Russian physicist, Lorentz had no other option than to probe Ehrenfest on the chances of survival of Lebedew's team. As to Solvay's request, it reminded him of the urgency of his succession in Leiden. He therefore cautiously asked Ehrenfest for details on his personal career, thanking him for sending his contribution to the *Encyclopedia*. He also praised Ehrenfest's paper on the ether, indicating that he had recognized in it the "*world of Boltzmann's ideas*[298]". This reference to Boltzmann was a real homage on Lorentz's part.

Ehrenfest replied[299] on April 24, 1912. He reported on the situation of the Moscow School of Physics, denouncing the brutal attitude of the regime which had caused the resignation of a hundred professors and associate professors.

Lorentz, who had been able to measure Ehrenfest's scientific value, was deeply struck by his moral qualities (his attachment to his adopted country, his empathy towards the victims of oppression)... However, these positive impressions didn't exempt him from taking the advice of two of his well-known German colleagues: Sommerfeld, and Ehrenfest's former mentor in Göttingen, Waldemar Voigt.

Sommerfeld replied in a more expansive manner than in his letter of February 25.

He confirmed his good opinion of Ehrenfest, indicating that it was shared by Einstein, who wished to welcome him as his successor in Prague (a wish which in Lorentz's eyes had the value of a recommendation).

But Lorentz remained cautious. He still waited for news from Brussels that would allow him to announce his intention to retire from his chair in Leiden.

A decisive event took place on May 12, when Solvay spoke to the members of the Belgian Chemical Society, announcing the creation of the International Institute for Physics. The following day Lorentz wrote to Ehrenfest[300] to mention his succession and to know the latter's intentions. He told him that the chair wasn't limited to Dutch candidates, and made his personal choice clear:

"*I thought of you because I have a high opinion of your work, and appreciate the depth, clarity and insight you demonstrate.*"

However, Lorentz took care not to find himself in the situation he had experienced with Einstein. He asked his candidate to let him know if he had any other position in mind[301].

Ehrenfest announced on May 19, 1912, that he accepted the Leiden offer. Lorentz immediately took steps to ensure the success of the operation. Notwithstanding his efforts, his succession was to be delayed for several months by red tape. Ehrenfest had to wait until the fall of 1912. His appointment at the University of Leiden, would have an ironical result. It would encourage Einstein to go to Holland regularly.

In fact, Lorentz's disappointment at having been deprived of his preferred successor would be alleviated in 1920 by Einstein's appointment as extraordinary visiting professor[302] at the University of Leiden.

Crowning of a career

Lorentz's new statute entered into force during the summer of 1912. Having become an extraordinary professor, he would give only one course on a theme of his choice.

In Haarlem, he had a laboratory at his disposal, and was assisted in his work by a qualified experimenter. He now could study the phenomena that captivated him, while devoting a great part of his time to the International Institute for Physics. Indeed, on May 17, he sent the following message to Solvay[303], confirming his agreement to the choice of May 1, 1912, as the official date of ISIP's foundation:

> "*The International Solvay Institute for Physics (I gladly accept this denomination), can therefore be considered as founded. I will announce it to my colleagues, and I will ask them to communicate their considerations of a more or less general nature on the best way to use the means that you wish to put at our disposal (...).*
>
> *I end this letter by telling you something which concerns me personally. It will be during next summer that I will leave the chair at our university that I hold up to now, by becoming an extraordinary professor. I will then live in Haarlem, close to the laboratory of the Teyler Foundation, of which I am the curator. This change will give me more leisure for my personal studies and duties, such as the one imposed on me by the presidency of the International Committee...*".

Lorentz's presidency of the ISC would extend well beyond the First World War. It would be the key to the success of the Physics Councils, giving them incomparable radiance. Lorentz would also benefit from ISIP's international fame. But there are two sides to every coin. Overloaded with work, he would be unable to continue his research on the photoelectric effect[304]. This would be done by Millikan in the United States, whose first results would be a signal for Lorentz to give up the game.

This observation raises a question: Why did Lorentz sacrifice his experiments for the benefit of an Institute, the mission of which would precisely be to evaluate and to support experimental projects?

Several reasons can be thought of. Lorentz may have been attracted by a position that would bring him in permanent contact with the elites of the profession, and keep him informed of the most recent results. We know that he cherished the opportunities that had brought him in contact with foreign colleagues. It is also quite possible that he wished to exercise a function in harmony with his experience, allowing him to promote new research in the most promising directions, and that he agreed to become the leader of an International Scientific Committee without realizing the burden of the task.

Still, by adhering to Solvay's project, Lorentz found himself in a position to advance physics and to put his talents at the service of an Institute, the action of which would stand until today as a model of international scientific cooperation.

We will recall Lorentz's relentless efforts during, and after, the Great War, to reconcile the physicists of the two camps, and to restore the scientific relations that had so brutally been interrupted. We shall witness his awareness of the universal character of science, an awareness he shared with Solvay and which enabled ISIP and ISIC to survive the disaster of the First World War.

Lorentz's lifelong commitment to science as a tool for peace was recognized in the 1920's, when he was asked, together with Marie Curie and with Einstein, to join the "Commission for Intellectual Cooperation", set up by the League of Nations, before chairing it in 1926.

It is worth emphasizing that the great turn in Lorentz's career – his resignation from his chair at the Netherlands' oldest University – resulted from independent proposals made by two private institutions: the Teyler Foundation and the International Solvay Institute for Physics.

One is entitled to wonder what would have happened if Lorentz had continued his experiments, and if he had devoted all his time to the study of the photoelectric effect. It may be that his indefectible commitment to ISIP, which he served during more than 15 years, deprived the world of an extraordinary event: the discovery of the photon by the champion of ether and the electromagnetic wave theory of light.

3.3 Second impact of the Council on Einstein's career: from Zurich to Berlin

Back to Einstein and to his career concerns. We saw that he chose to settle in Zurich, despite the offers that were made to him, in particular by Warburg, president of the Physikalisch-Technische Reichsanstalt in Charlottenburg, and by the University of Vienna, on Hasenöhrl's proposal. Yet, Einstein's presence at the ETH was short-lived. A politico-scientific operation, led by first rank physicists, soon managed to lure him away from Switzerland, and from his family. Einstein settled in Berlin from the spring of 1914, and kept his post there for almost twenty years.

Einstein's first stay in Berlin

Shortly after the Brussels meeting, Einstein re-examined the question of specific heats. Aware of Nernst's displeasure[305] with the dispute about the status of the heat theorem, he knew that his quarrel with the irascible chemist had been amplified by his rejection of Rubens' interpretation of data which had served as a basis of the Nernst–Lindemann formula[306]. Seeking to defuse the conflict, Einstein decided to go to Berlin. A stay in the German capital seemed appropriate for several reasons.

On the one hand, Einstein wanted to give a new impetus to his collaboration with Warburg on photochemical decompositions. On the other hand, he felt that a renewed contact with Planck and Nernst, would allow him to iron out their disagreements on quantum theory and on its consequences.

Einstein also planned to visit Fritz Haber, a chemist he had met during his stay in Karlsruhe of September 1911, and who was about to take over the direction of the Kaiser Wilhelm Institute of Physical Chemistry and Electrochemistry. Last but not least, his stay would provide an opportunity to meet the astronomer Erwin Freundlich, his German correspondent in the study of gravitational effects on light.

The Berlin intermezzo took place in April 1912, during the Easter holidays.

Haber was delighted to see Einstein again, and to meet his wife, Mileva Maric[307]. He regretted not being able to receive the couple at his home, and set about finding accommodation for them.

But that wasn't all. Einstein's stay in Berlin had a consequence of quite another nature. His reunion with his cousin, Elsa Löwenthal, marked the beginning of an affair that would spell the end of his marriage, and would eventually separate him from his sons.

Back from his trip, Einstein sent this message[308] to Elsa:

"I can't even begin to tell you how fond I have become of you during these few days.

And I will also come to see you before too long (I think at the end of this semester), if you feel this is all right. It is such a pity that we don't live in the same town. The chances of my getting a call to Berlin are, unfortunately, rather slight, as I must admit to myself when I think about it clearly...".

What Einstein didn't know, was that two influential Berliners, Haber and Planck, were already forging plans to attract him to the capital of the German Reich[309].

The call from Berlin

Haber took action in January 1913. Strongly impressed by Einstein's genius, he reckoned that a close collaboration with this extraordinary man would lead to a solution of the quantum enigma. This ambitious goal seemed all the more attainable since his position at the Kaiser Wilhelm Institute enabled him to recruit Einstein.

By doing so, Haber would respond to the wish of Hugo A. Krüss, an official of the Prussian Ministry of Education, who worked under the direction of Minister Friedrich Schmidt-Ott.

Krüss was the author of a *"Memorandum"* which preached the support of German scientific research through public and private funds. He belonged to the group which in 1911 created the Kaiser Wilhelm Society (KWG) for the promotion of sciences. One of the Society's objectives was the foundation of Research Institutes. Another was the recruiting of seasoned scientists, capable of leading them[310].

Here is an excerpt of a letter[311] that Haber sent to Krüss on January 13, 1913:

> "In a conversation about the Ord. Professor of Theoretical Physics at the Polytechnical Institute in Zurich, Dr. Albert Einstein, which we had in the year that just ended, you brought up the question whether there could not be created position fort this extraordinary man at the Institute in my charge.
>
> After having turned this idea over in my mind for quite some time, I have become convinced that the realization of this idea would be of the greatest advantage for the Institute, and that, from the personal side, it could probably be attempted with some chance of success. Even though I did not go so far as to give Mr. Einstein any hint of it, I did find out that, completely absorbed by his investigations as he is, he would gladly do without the large course of lectures that he is obliged to give. Further, I ascertained that he has no fundamental misgivings regarding Berlin.
>
> It is true that he declined an invitation to join the Reichsanstalt, which Mr. Warburg extended to him some time ago, but it is precisely the reasons that led him to this decision which give me hope that, in principle, he would not react negatively to an invitation from our Board of Trustees.
>
> I discussed the idea of getting this man with Privy Councillor Koppel. As you surely remember, Director Schmidt also took an interest in this idea. Director Schmidt aptly remarked that it does not seem necessary to establish an institute for him because he is not an experimentalist. On the other hand, it is precisely this circumstance that makes it easy to incorporate him into an existing institute. Even a theoretical physicist of his orientation is in need of certain resources in order to study experimentally some topic or other from time to time, or to have it studied by an assistant or co-worker. I could grant him the space and the equipment for it on the upper floor of the Kaiser Wilhelm Institute of Physical Chemistry without thereby impairing the normal work of the Institute. Mr. Privy Counc. Koppel was in principle ready to request the recruiting of Einstein...".

Haber also felt confident that *Geheimer Kommerzienrat* Leopold Koppel[312] would agree to provide the necessary funds to offer Einstein a good salary. Unaware of Planck's intentions to attract Einstein to Berlin, he told Krüss that the author of the quantum theory was Einstein's closest colleague in Berlin and that he should be informed of the project.

Planck's assets were more powerful than Haber's. He was Secretary of the Physico-Mathematical Class of the Prussian Academy of Sciences, and would soon become President of the University of Berlin

Planck also knew of Einstein's desire to be released from the obligation to teach. He was aware of the attraction that a seat in the Academy would represent in his eyes.

Now it happened that a special Academy chair was vacant since the death in March 1911 of Jacobus van 't Hoff, first Nobel Prize winner in chemistry[313].

So, all Planck had to do was to find the extra money that would increase his chances of success. This was the mission he entrusted to Nernst, his good colleague who had close contacts with the Berlin business community, and in particular with Koppel.

The sequel is well known. Nernst had no difficulty in winning Koppel's support[314].

A proposal to elect Einstein to the Academy was submitted on June 12, 1913. The letter bore the signature of the four Berliners who had taken part in the Brussels Council of 1911: Planck, Nernst, Warburg and Rubens.

A few weeks later, Planck and Nernst went to Zurich to inform Einstein of the project.

The ETH-professor hesitated for a moment, but finally accepted the proposal. On July 14, 1913, he sent this message[315] to his cousin:

Dear Elsa,

I have not answered your letter. But in its place, I will now tell you some really good news. A few days ago, I was visited by Planck and Nernst, who insisted on coming from Berlin for the sole purpose of offering me a position at the Academy. Thus, next spring at the latest, I'll come to Berlin for good. It is a colossal honour that has thus been bestowed upon me, since I will become Van 't Hoff's successor.

I already rejoice at the pleasant times we will spend together. But don't tell a soul a thing about the matter. A decision of the plenum of the Academy is still needed, and it would look bad if something were to become public knowledge...

The official announcement was made on November 22, 1913. Einstein learned that his election to the academy had been approved by the Kaiser. He was told that his annual salary (900 marks) would be increased by a sum of 12 000 marks.

Haber hadn't won his case. However, he had reasons to be satisfied: Einstein would occupy an office in his Berlin-Dahlem Institute upon his arrival in Berlin, and would settle in a housing near the Institute[316].

Before concluding this chapter, it is worth mentioning one of the long-term effects of the "*Witches' Sabbath*" on Einstein's academic career.

Indeed, it is known that Lindemann, Nernst's British assistant who had taken part in the Physics Council, had been struck by Einstein's personality (an impression he shared with his father[317]). Having taken the lead of the Clarendon laboratory in 1919, he awarded Einstein an honorary doctorate from the University of Oxford. Lindemann completed the distinction by conferring Einstein the title of *Fellow of Christ Church College*, and by offering him for five years an annual scholarship of £ 400[318]. Einstein benefitted from the advantage from 1931 to 1933.

It should be added that Lindemann, the later Lord Cherwell, took care of recruiting Jewish physicists expelled from Germany in 1933, and that he became one of Winston Churchill's main scientific advisers during World War II.

Chapter 4
Foundation of the International Solvay Institute for Physics

Several facts, reported in the previous chapters, are indisputable indications that Solvay had made up his mind before the end of the Council: he would support experimental research on specific topics, and would do so with Lorentz's help.

As to the Solvay subsidies, they would be granted by an International Institute.

We shall now see that this Institute took shape after just a few months, thanks to Solvay's determination, to Lorentz's efficiency, and to the perfect agreement between these two exceptional representatives of complementary worlds.

4.1 The Lorentz proposal

Let us go back to the lunch of November 4, 1911. It is likely that Solvay reminded his guest of his wish to support new experimental research on radioactive phenomena and on Brownian motion. This last topic happened to be the subject of the doctorate of Lorentz's daughter, G. L. de Haas-Lorentz, which she prepared under her father's direction (she would defend her thesis on September 24, 1912). We may therefore assume that the idea interested Lorentz, and that Solvay's wish was welcomed with enthusiasm. However, it is more than probable that Lorentz convinced Solvay to extend the idea to "*experiments of which the need had been established by the Council's deliberations*[319]". In other words, they agreed on the necessity of an International Institute, the task of which would be to implement an ambitious subsidy program.

A few days later, Solvay went to Leiden to visit Kamerlingh Onnes' cryogenic laboratory. This trip enabled him to remind Lorentz of his foundation project, and to specify his goals:

  i. To promote research in physics by granting subsidies to qualified experimenters of all nationalities.

ii. To encourage the scientific movement in his country by granting scholarships to young Belgian researchers who wished to improve their capacities by spending some time abroad.

Solvay used the opportunity to present Ostwald's "*Memoir*[320]" on the creation of an International Institute for Chemistry, explaining that he hadn't found the time to study it. Would Lorentz be so kind as to examine the document?

Lorentz accepted the request. He recorded the result of his reflections in a long letter of January 4, 1912[321]. His first observation was that Ostwald's proposal didn't respond to Solvay's intentions: it didn't contain any provision for the support of exploratory research. The planned Institute would only provide a variety of services, such as chemical data collection, the standardization of current formulas, the creation of a uniform nomenclature and the setting up of a central library for the entire chemical literature[322].

Lorentz proposed an Institute designed to meet Solvay's objectives. Its scientific management would be entrusted to an "International Scientific Committee" (ISC), whose composition would evolve in time. This ISC could be, at the beginning, the Select Committee that Lorentz had constituted during the Council.

A local "Administrative Commission" (AC) would be responsible for financial and administrative matters.

The Institute would serve two purposes:
— The granting of subsidies to experimenters from all countries for research on issues determined by the ISC.
— The granting of scholarships to Belgian researchers on proposal of the AC.

In case of agreement from Solvay, Lorentz would submit his project to the members of the Select Committee in order to obtain their adhesion. The ISC being constituted, the Institute would be able to start its activity without any delay. Lorentz took care to underline this advantage in his letter to Solvay:

"If you agree to entrust the task of the first scientific committee to the one that already exists, we could start working as soon as you would have taken your decision. I am telling you this because I remember that you expressed yourself on this matter during our conversations in Brussels...".

The report ended as follows:

"This is, very honoured Mr. Solvay, what I thought I should submit to you. I would be very happy if I managed to seize your intentions and to offer you a scheme that you could approve... I do not have the pretense of saying that what I suggest to you is the best thing you can do, but I am firmly convinced that by acting in the indicated manner you will do something very useful: very helpful for science and very important for your homeland, to which your Foundation will do great honour...".

Solvay reacted with enthusiasm. He realized that the Council had put him in touch with a providential man: a first rank theoretician, whose intelligence and tact had been greeted by everyone... and who practiced four languages. Above all, he appreciated the Council chairman's internationalism, a laudable commitment founded on an acute consciousness of the universal role of science.

In short, Solvay understood that he had found the ideal scientific director of the planned Institute. These were his words in his answer[323] to Lorentz:

> "As you understood, and as you say in your letter, I'm not only aiming at the progress of science in general; my ambition is to give impetus to the scientific movement in my country. I already worked hard in this direction, either by founding institutes, or by other interventions. And now it is a new effort that I still want to make. I thank you for wanting to help me with your advice: no one is more authorized in this respect than you. Your project pleases me because, in its outline, it responds to what my life's ambition has been...".

However, despite his enthusiasm for the Lorentz proposal, the industrialist knew that he couldn't commit himself without taking the advice of his scientific collaborators. He consulted Héger and Hostelet. The director of the Physiology Institute declared himself very much in favour of Lorentz's project[324]. But his opinion wasn't shared by Hostelet, who declared in a letter[325] to Solvay:

> "If I understood your intentions correctly, you want to promote classical science through your foundation in a more novel way than that envisaged by Mr. Lorentz (...). I do not find in Mr. Lorentz's project the essence of the idea of the Physics Council that you wish to extend and to develop[326]. I do not recognize any of the characteristic provisions to which its success is due, that is to say a happy meeting of individual initiatives, and at the same time, the systematic elimination of undesired and diverting interventions of official and academic authorities. In fact, what you want most is to intervene at the right time to give the necessary impetus, even it means quitting then to wait, while leaving others full freedom..".

Solvay's adviser for physics put his finger on a point that hadn't been foreseen in the Lorentz proposal: ISIP's duty to organize regular Council meetings, similar to the one of 1911. This point didn't escape Solvay's attention. Confronted with Hostelet's other critical remarks, he realized that the acceptance of the proposal in its entirety would deprive him from any initiative in the future. Another concern was Lorentz's idea that the ISC members should be appointed on proposal of the International Association of Academies. This would clearly be a source of unacceptable academic interference in the Institute's affairs. Reflecting on these questions, Solvay decided to exercise control over the drafting of ISIP's statutes. He asked Héger to go to Leiden and to prepare a preliminary draft in consultation with Lorentz.

4.2 The Institute's Statutes

The meeting in Leiden took place on January 20, 1912. Solvay's emissary had been asked to ensure that three conditions were met:
- ISIP would have its headquarters in Brussels, where appropriate rooms had been reserved in the Institute for Physiology.
- It would organize regular meetings of the Physics Council on the model of the 1911 meeting.
- ISIP's statutes should include an *impartiality clause*, intended to limit the undesired and diverting influence of academic circles[327].

Lorentz willingly subscribed to these conditions. He proposed the introduction of the following impartiality clause[328]:

> "*The founder wants above all the Institute to demonstrate in all its actions a perfect impartiality, to encourage research undertaken in a true scientific spirit, whether it is carried out in an isolated laboratory or in a large scientific centre. He considers it desirable that this trend be reflected in the composition of an International Scientific Committee. Therefore, if there are scholars who, without occupying a high official position, can be considered for their talents and personality as worthy representatives of science, they should not be forgotten by those who nominate candidates for vacant places.*"

Solvay approved Lorentz's formulation of the clause, but he asked[329] to add a few leading lines "*in the sense of an objective interpretation of the phenomena*". The clause thus became:

> "*The founder wants above all the Institute to demonstrate in all its actions a perfect impartiality, to encourage researches undertaken in a true scientific spirit, whether they are carried out in an isolated laboratory or in a large scientific centre,* **and especially since, at equal value, they will have a more objective and therefore a more satisfactory character.** *He considers it desirable that this trend be reflected in the composition of an International Scientific Committee...*".

Solvay also wished to be kept informed of any discussions that would take place during the meetings of the Physics Council. He wanted Herzen and Hostelet to attend these meetings, and asked for a clause to be added to this effect (see section 2.9): "*The Administrative Commission reserves the right to nominate two representatives among Mr. Solvay's collaborators who will be invited to attend all meeting sessions*".

Brillouin's influence

In his answer to Solvay of February 2, 1912, Lorentz said: "*The benefits of your foundation must be accessible to all who deserve it. Isolated and little-known talents should not be excluded...*".

This statement, totally consistent with the industrialist's wishes, echoed some remarks made by Brillouin in his letters to Lorentz on January 27, 28 and 29, 1912. They tell us that the Council's chairman had asked Brillouin for advice on certain questions regarding the new Institute (Lorentz's letters have unfortunately been lost, but we have Brillouin's answers[330]). These are the latter's words in his reply to Lorentz of January 27:

> "I need to think about your letter. I will write to you shortly about the matter. PS: The foundation is a scientific one. The scientific competence belongs to you, in particular, and to the Council[331] constituted in Brussels last November – not to Mr. Solvay. That solves the question."

We deduce from the postscript that Lorentz had inquired about Brillouin's opinion on ISIP's scientific management.

In a much longer letter of January 28, Brillouin exposed his views on the composition of the International Scientific Committee:

> "I naturally thought a bit about the Solvay Foundation and about the composition of the future Committee. The first result of my reflections, once the pleasure of the meeting had passed, was to confirm myself in the impression that I communicated to Mr. Solvay in June 1911: that I should not be part of this Committee. In two letters to Mr. Solvay, I apologized for being unable to understand spoken English and German, and for having to be a very inactive member of the Congress. Reading the report of the sessions confirmed me in my regret. For a Committee, which has decisions to take, this is an impossible situation. Already in November at our little preparatory meeting, I noticed that I didn't understand anything (underlined) of what Mr. Warburg and Mr. Nernst said, and only little of what Mr. Rutherford said. It is quite impossible to be a useful and conscious member of a meeting under such conditions. You must therefore choose another French member. I add that, being neither a mathematician nor an experimenter, I feel very poorly qualified to represent French experimenters in the meeting. Here is my first resolution.

> This being said, I wonder if the committee that was improvised in November is not a little too restricted. It naturally depends on the nature of the foundation it will have to manage, or of which it will be the scientific boss.

> Yet it seems to me that J. J. Thomson, Lord Rayleigh, Ramsay in the case of physical chemistry, and perhaps others, cannot all be left out of the first grouping. Likewise, I wonder if in Germany Messrs. Warburg and Nernst, both from Berlin, shouldn't be accompanied by one or two representatives from other parts of the empire. I imagine that there are also rivalries among them, and that it isn't Mr. Solvay's intention to restrict his generosity to a small group.

> In my own country, I have lost many illusions during the last ten years. To be really useful, the Solvay Foundation must not fall prey to any

exclusive group. This is where you will recognize my sorrowful spirit and my inability to find solutions, due to which I am not part in France of any commission of any kind. Our Academy is rich, and the prizes and subsidies it distributes are numerous. The Solvay Foundation shouldn't be an annex to it. We shouldn't therefore rely on Academy Secretaries, but carefully choose among members of the physics section. There is one of them, the youngest, with a rare experimental skill and a completely independent spirit, who isn't a professor anywhere (underlined), *but who works at the École Normale's laboratory, living on the small remains that his parents left him: it's Mr. Villard*[332]. *The others teach mostly in rich establishments, to which many endowments are already going.*

You know, on the other hand, what an unfortunate role the appalling centralization in Paris plays in my country. It seems to me that it would be very useful to call on a provincial researcher (underlined), *for example the university professor Mr. Fabry*[333], *a former pupil of the Ecole Polytechnique, whose spectroscopic works, which started under the influence of my old comrade and friend, Mr. Macé de Lépinay*[334], *have been pursued with method and precision despite the changes of collaborators.*

When you think of a chemist, it seems to me that my friend from Toulouse, Paul Sabatier, who refused to move to Paris after Moissan's death, would be perfectly suited[335].

But this is much more than what is needed, and you may wonder why I don't write the names that you are thinking of. My concern is that the Solvay Foundation shouldn't go to organizations and establishments that are already relatively wealthy, but to those which it will actually serve. Furthermore that the French delegates be those who have no scientific egoism and do not form a tightly-knit chapel – whatever the importance of their personal research, but whose technical skill, astute observation and precision of measurements, isn't contested by anyone, and by whom all young French physicists feel well represented.

I believe that one or other of the names I gave you would be very well received, but I don't wish to interfere with matters that no longer concern me after having recognized my own inability to be part of the Committee. It is only for guidance, and perhaps outside the direction you intend to give to the Solvay Foundation. It is difficult to find solutions that satisfy (scientific) Europe and one's own country. At home we may not always be free enough from rivalries, jealousies, prejudices, either from school or from career.

Also, I beg you to look at this letter only as a starting point for your reflections, for your information, and to take no account of it if the overall view you have of the foundation project, and that your European scientific relations allow you to assess, lead you to decisions unrelated to my writings...".

It is likely that Brillouin's advice weighed in the balance, insofar as it agreed with the founder's position and with Lorentz's own experience. It may have contributed to the taking into account of Solvay's demand of an "impartiality clause".

Brillouin's letter was also important for other reasons. Apart from being a first-hand testimony of the scientific situation in France at the beginning of the twentieth century, it dramatically emphasized the difficulty for most Council members to follow a scientific debate in three languages (a point we already underlined in our foreword).

Yet, some of Brillouin's statements should be taken with caution, such as his remark about tightly closed "scientific chapels" which clearly referred to Marie Curie and to her Institute, created in 1909. They confirm the impression which emerges from the extract of his letter to Lorentz of January 29, which we already quoted in section 2.9.

Indeed, answering in this letter Lorentz's question about the consequences of the Curie-Langevin affair, and whether Mrs. Curie should remain part of ISIP's Scientific Committee, Brillouin summarized his views as follows:

1 Mrs. Curie represents the discovery of radium and the in-depth study of many of its properties; a colossal achievement, superior to that of most physicists of this century, of inappreciable consequences.

2 She doesn't represent French physicists – few will choose her as an intermediary for their application to the Solvay Foundation, especially among those who study radioactivity outside her laboratory, and who are not part of the narrow chapel that surrounds her.

Whatever the value of Brillouin's opinion, it is well known that Marie Curie met a resounding failure in her candidacy for the Academy of Sciences. Now it happens that Brillouin was a competing candidate at the time, and that he was the one who won the fewest votes... This might be the source of his mistrust of Academies and of the bitterness of his reaction to Lorentz's inquiry. It might also explain his judgement on Marie Curie's "lack of representativeness".

Fortunately, Lorentz chose to ignore Brillouin's arguments against Marie Curie. On the contrary, he made sure that history's first laureate of two Nobel Prizes kept her place in ISIP's Scientific Committee. Not only did Marie attend seven Councils (those that took place under Lorentz's chairmanship) she also served ISIP's purposes for twenty years (from 1912 to 1932) devoting part of her precious time to the evaluation of the numerous research projects[336] submitted to the Institute during its first financial years.

Let us also recall that Lorentz refused to follow up on Brillouin's wish to withdraw from the ISC, and that he recognized the relevance of his fears that ISIP might be tempted to favour physicists attached to well-known laboratories... All in all, the advice he received from this French physicist, dear to Solvay, reminded him of the tensions specific to the academic world, and of the difficulties he would have to overcome during his chairmanship.

Lifespan of the Institute

On February 16, 1912, Lorentz received a letter[337] from Héger, informing him of the lifespan Solvay intended to grant to his foundation:

> "*Up to now, Mr. Solvay only made foundations of a limited duration. Twenty-five or thirty years seemed long enough to him to produce a change in the overall situation, especially in scientific matters. Mr. Solvay also believes that it is the role of the State to subsidize and to organize scientific institutions, and he hopes that the State will fulfil this task in thirty years better than it does today. He agrees with his customary generosity to devote 1 million to the goal summarized in the Institute's statutes; but he will give this million to a Bank or to an Insurance Company, so as to provide an annual interest to the Foundation. I understand that 1 million, a non-repayable grant for twenty year's brings at least 75,000 francs per year. Mr. Solvay is asking if you would prefer a higher income during a shorter time – or a lower income for a longer time, thirty or even forty years... Please give some thoughts to this original foundation form, and let me know your feelings, or share them directly with Mr. Solvay...*".

Lorentz, who favoured a lifespan of fifty years, responded[338] in an attempt to justify this longer term:

> *Dear and Honoured Sir,*
>
> *Mr. Héger tells me that you may prefer to create a foundation of limited duration, of twenty-five or thirty years for example. This is a point on which it is difficult for me to express an opinion, given the impossibility of predicting what will be the look of the world in about fifty years. However, I fully understand your thought. Indeed, we can hope that governments will realize more and more the importance of scientific research, and that we will eventually witness an organization independent of the individual efforts of private persons.*
>
> *In short, it seems to me that you must make a decision in this matter which is most in accordance with your ideas. I very much hope that the approach you are going to take will bear the imprint of your personal views. Only, I think I should ask you to grant the Foundation a life of fifty or at least forty years; I haven't the slightest doubt that she could render great services during this period of time.*
>
> *On the other hand, if after all you want to create it in perpetuity, it would be better to provide for an indefinite increase in capital. According to article 13 of the preliminary draft statutes, a tenth of the annual income will be capitalized. We could specify that this will only be done as long as the capital doesn't exceed a certain limit...*

I take this opportunity to tell you about a news that I recently got from Russia, and which is very apt to show the great utility that the new Foundation could have.

You probably know that about a year ago the Russian government took an attitude towards the University of Moscow which caused a large number of professors to resign. Among them is Mr. Lebedew[339], one of the best physicists of this time, who supervised the work of eighteen people in his laboratory. All these physicists were forced to leave the premises and Mr. Lebedew was glad to find refuge for himself and for fourteen of his collaborators in a laboratory which has been set up in a private house, and which belongs to the Municipal University of Moscow (called People's University). This solution could be found thanks to the generosity of a certain number of enlightened men, but the resources of the new laboratory appear to be very limited.

Mr. Lebedew is a very skilled experimenter and has been noticed for his important and in-depth research. This is how he succeeded (like Righi in Bologna) in producing electromagnetic waves similar to those discovered by Hertz, but of a much smaller wavelength. Mr. Lebedew managed to reach less than a centimetre; this allowed him to observe with these rays, invisible all (or almost all), the phenomena presented by light rays, such as interference, refraction, double refraction, and polarization.

Mr. Lebedew's research has been continued since 1895, and he now proposes to undertake a systematic study of the absorption and refraction of these electromagnetic rays in a wide variety of substances. This will be very important for our knowledge of the constitution of matter, and it would be regrettable if the experience gathered by Mr. Lebedew in this area was lost... I also wish to recall that he has been the first to experimentally demonstrate the existence of the pressure exerted by the rays of light, a pressure which had been predicted by Maxwell and which plays a considerable role in the theory of radiation and in that of cosmic phenomena. Another question Mr. Lebedew addressed is whether the rotation of a body (without electrical charge) can produce magnetic effects.

I allowed myself to tell you all this because the example shows that the Foundation could immediately do something very useful. The difficulties encountered by Mr. Lebedew also make me think that we should not count on a very rapid progress.

I am firmly convinced that in the struggle against ignorance and obscurity truth and light will prevail, but I think that we should be very optimistic to believe that in a country like Russia the future that we dream of will be realized in half a century...

Yet, despite the strength of his argument, Lorentz had to face the obvious: the annual income of 1 million, as a non-repayable grant for fifty years, didn't make it possible to cover the operating costs of ISIP and the granting of subsidies (first reason of being of the new foundation). Forced to reconsider his position, he told[340] Solvay:

> "As you left me the choice of the number of years for which the Institute for Physics will be founded, I believe I must propose you to fix its lifespan at thirty years."

It meant that ISIP's annual income would be of the order of 54 000 francs. One third of it would be spent on subsidies, another third to scholarships, and a final third to the operating costs (including 1 000 francs[341] for the Scientific Committee's president and 500 francs for its secretary).

4.3 Constitution of the International Scientific Committee (ISC)

A particular event, already mentioned in section 3.2, produced a sudden acceleration in the Institute's foundation process. It led to the following message from Solvay to Lorentz (letter[342] of April 19, 1912):

> "The Belgian Chemical Society will celebrate its 25th anniversary on May 12, 1912, and I am asked to present a talk on this occasion. Having naturally little to say as a chemist, I thought I might announce, in agreement with the president, the foundation of the International Institute for Physics, and that while pointing out some statutory peculiarities. I must prepare my talk right now...".

Lorentz realized at once that ISIP's founding was imminent. He immediately wrote to Ehrenfest to accelerate his succession (letter of April 20), and took care to reassure Solvay (letter[343] of April 21):

> "I sent several copies of the preliminary draft statutes to my colleagues a few days ago, and I expect to receive their answers this week. Besides, I have no doubt that they approve the general ideas which are the basis of these statutes. It will be only in the details that we may propose some change, and you can certainly prepare for the talk that you will present at the Chemical Society. Just like you, I think that the Society's meeting offers you a good opportunity to say something about the new Institute and to highlight its significance. You probably have a copy of the draft in the last form we gave it...".

In anticipation of his speech to the chemists of May 12, Solvay reformulated article 2 of ISIP's statutes:

> Article 2: *The Institute's goal is to encourage research which is likely to extend and, above all, to deepen knowledge of natural phenomena in which Mr. Solvay is constantly interested. The Institute focuses on the progress of physics* **and of physical chemistry** *without, however,*

Foundation of the International Solvay Institute for Physics   123

*excluding problems belonging to other branches of the natural sciences, provided, of course, that these problems relate to physics **or to physical chemistry**.*

Lorentz, on his part, was active to obtain his colleagues' adhesion to the draft statutes. He asked them to take a seat in the Institute's Scientific Committee, and requested Knudsen to accept the post of secretary.

Lorentz's consultations had a prompt effect: on May 1, 1912, he was able to announce the acceptance of all members in a letter[344] to Solvay. It contained the encouraging reactions of five ISC members:

- Rutherford: *"I highly appreciate Mr. Solvay's great generosity, and the interest he takes in the progress of scientific knowledge. I think that his Foundation will be very useful to science in general if the International Council takes a high point of view"*.
- Nernst: *"The statutes seem to me very well designed; the general trend and all the details are happily expressed"*.
- Warburg: *"I find it excellent that, in accordance with Mr. Solvay's wishes, the Institute will only be founded for a limited number of years"*.
- Brillouin: *"I am pleased to convey to you my complete approval of the organization to which you have arrived, and my gratitude towards Mr. Solvay for his generous initiative"*.
- Kamerlingh Onnes: *"I have the pleasure to return to you the draft statutes, which I find excellent; I don't have to suggest any modification"*.

4.4 Birth of the International Institute for Physics

Solvay didn't hide his satisfaction at reading the above comments. He told[345] Lorentz:

"I am very pleased to know that the draft statutes of the International Institute for Physics received the approval of your colleagues of the Scientific Committee, Mrs. Curie, however, having not yet replied...I must now find an agreement with the King, the City of Brussels and the University. However, since the premises at the Institute of Physiology is at stake, and since there are decisions to be taken in this matter, which will take a lot of time, I would suggest, if you agree, to consider the Institute for Physics as established now, and functioning provisionally without Administrative Commission. I will immediately put the funds necessary for this purpose at your disposal, a first annuity or what you would indicate to me...".

Thanks to Lorentz's diligence, Solvay would be able to announce ISIP's birth at the meeting of the Belgian Chemical Society. The Institute surprised by its novelties, compared to Solvay's earlier foundations. It didn't involve any new building, nor did it require special tools, or staff with fixed commitment. Its headquarters were located in Brussels, but the members of its Scientific Committee were scattered throughout the world.

The Institute's mission was to stimulate progress in physics and in physical chemistry by granting subsidies to isolated researchers who would be encouraged to continue their work independently. As to the "Physics Council" which had met in 1911, it would be called upon to periodically hold new meetings, in order to take stock of the latest developments, to identify the most serious problems and to open avenues for their solution.

ISIP's neutrality was exemplary. Its Administrative Commission was headquartered in Brussels, capital of a neutral country; the ISC chairman resided in the Netherlands, another neutral country; its secretary resided in Denmark, a third neutral country. Something never seen before.

Evolution of the founder's position

ISIP's total freedom of action was quite surprising, since physics had always been Solvay's main preoccupation. We saw that his desire to meet new challenges had even been increased by the Council's meeting (see Solvay's closing speech of November 3, 1911, his Notes[346] and our comments at the end of section 2.9).

What should we conclude from this?
  i. That Solvay had by no means renounced his scientific ambitions, and that he intended to pursue the building of his *Gravito-Materialitic* theory independently of the opinions and concepts of professional physicists.
  ii. That he wished to submit his ideas to experimental tests, especially his views on the origin of radioactivity, and that he welcomed the possibility to call on the expertise[347] of particular ISC members.
  iii. That for this isolated thinker, dissatisfied with current physical theories, there was no contradiction between the pursuit of personal research and the support of new experiments that might provide additional evidence of the need of an alternative approach.
  iv. That the ISC was an exceptionally qualified scientific body, capable of designing such experiments, and of selecting more general projects which deserved to be supported by a Solvay-subsidy.

The fact that Solvay initially intended to call on ISIP's scientific committee to boost his own research is clearly suggested by the end of his closure speech:

"*You, specialists, could easily, I believe, help me to obtain satisfaction on this matter, by resolving the question* (of the origin of radioactivity and of Brownian motion) *one way or the other,* **under my control**...".

That this wish remained on his mind until the end of March 1912, emerges from his message[348] to Héger on March 27, 1912:

"I intend, with the help of grants, to have verifying research carried out on my principles by specialists from everywhere, and I still firmly believe that by doing the necessary with difficulty, I will find myself being the theoretician capable of solving correctly big open problems. This subsidy plan is, as you know, not new in my mind, and has also been adopted by the Physics Institute, which proves it to be good...".

However, Solvay's attitude suddenly evolved during the spring of 1912. His change of mind is testified by a letter[349] to Lorentz of May 20:

"I see that there seems to be a perfect community of views between us... I also wish to state that I wouldn't like the Institute to bear the imprint of my active personality... Here in Brussels, my two University Institutes (Physiology and School of Commerce) are clearly separated from my personal Institutes (Physiology and Sociology); I have nothing to say, nor to see, in the first two, and the University has nothing to say, nor to see, in the last two, I so wanted it. The International Institute will be roughly equivalent to the first two; it has moreover statutes which characterize it...".

Indeed, ISIP's founder had accepted the statutory provisions which guaranteed its scientific independence... Hence, why this need to clarify his position?

Solvay's move seems to have been caused by the joint influence of Héger and Tassel, his oldest advisers.

We know from a letter of May 13, sent to Lorentz by Solvay's secretary, Charles Lefébure[350], that Tassel had been asked to take part in the "implementation phase" of ISIP's founding project. Now, it happens that Tassel and Héger, two retired university professors, approved the total freedom[351] granted to ISIP's Scientific Committee. It is therefore probable that their opinions weighed on Solvay's decision not to intervene in the action of the new Institute. This seems to be confirmed by Héger's statement in a letter[352] to Tassel of May 21:

"The solution is getting closer, and it is fully compliant with our hopes. Please tell Mr. Solvay how strongly I congratulate him...".

Whatever the reason that Solvay changed his mind, his statement *"I will have nothing to say, nor to see"* in ISIP's management, was a noteworthy development.

It meant that the ISC, an independent scientific committee, would be totally free to choose beneficiaries of private research subsidies: a world premiere in scientific patronage.

Solvay's resolution of May 20 was all the more commendable since he was impatient to have his views verified by experiment, in particular those regarding the origin of radioactive phenomena. This is a letter[353] which he sent to Marie Curie in the aftermath of the Brussels meeting:

> *Madam Curie,*
>
> *I am very happy to know that you are willing to investigate in the way that I have in mind, considering that the matter offers a real interest, which rather reassures me about the time that it will snatch from your regular studies.*
>
> *It is clear that your initiative should remain intact as to the choice of the experiments to be carried out. I wouldn't want to undermine that at any price, but the more the experiments will be designed to obtain a clear-cut result, the happier I will be, since my interpretation of the fundamental phenomenon of radioactivity is also very clear. I notice with pleasure that you intend to use a great quantity of radium; this seems to me desirable, for I have believed for a long time that if we had been able to dispose of 1 kilogram of radium at the beginning, the idea would never have occurred to interpret the origin of its energy as one did. This quantity would have behaved, in my opinion, quite differently from what is observed with the small quantities that are commonly handled.*
>
> *I am far from being a man informed on the phenomena of radioactivity: I suggest lead to surround the possible mixture of silver chloride and radium because I imagine (to see if it's true) that it is the best screen capable of preventing radium from feeding on the radioactive rays that always exist scattered in the external environment, and which most likely come directly or indirectly from the sun. To do it right, before placing the lead around the mixture of radium and silver chloride, it should probably be freed from radioactive rays which impregnate it, for example by immersing it for a while in silver chloride. Of course, you must have standard methods for all of this, and it is quite possible that I am expressing naive thoughts.*
>
> *There is perhaps a simpler experiment than the previous one to perform with lead. Here it is, with the reasoning which seems to justify it (...).*
>
> *When you say "We can try to see if there is a difference, however slight, which might occur in the radioactivity and heat release of radium as a result of the presence of absorbing substance", it is perhaps an experiment of the above kind that you yourself have in mind. It goes without saying that my grant wouldn't be limited to the amount you received, if it would be necessary to succeed*[354]...

The letter indicates that Marie Curie showed some interest in Solvay's concerns, despite the general consensus in 1911 that radioactivity wasn't due to the external environment. It also tells us something about the way in which Solvay felt authorized to ask ISC-members to check some of his own predictions (see also section 6.5). ISIP's founder conformed, to some extent, to the scheme he had adopted with his earlier foundations: the creation, on the one hand, of an independent Institute, directed by professionals, and the pursuit, on the other hand of personal investigations with the help of private collaborators. The separation he intended to maintain between these two components didn't exclude, in ISIP's case, the possibility to call on ISC-members to carry out experiments for his own benefit, with the clear understanding that none of the costs would be covered by the Institute.

Announcement of ISIP's birth

Let us now focus on Solvay's speech at the meeting of the Belgian Physical Society. The title was: "Contribution to the study of the constitution of matter, of the ether, and of energy".

> "*I imagine, said Solvay, that by appointing me an honorary member of your Society, you have aimed above all at the industrial whose obstinate labour contributed to the success of a now prosperous industry, but the beginnings of which were fraught with obstacles and daunting difficulties. So, there are some of you who have certainly been astonished to learn that for the first time I have the honour to speak to you, I took as subject the study of the constitution of matter and energy...*".

Solvay recalled that in 1858 he believed he had discovered by pure logical deduction a property of specific heats, known under the name of "law of Dulong and Petit". He also mentioned that long before Linde he had been absorbed by the problem of the liquefaction of gases, and by the production of low temperatures.

> "*But, he continued, these were only secondary expressions of my scientific work. This has been dominated for almost thirty years by a great philosophical concern: that of finding a simple interpretation of the science of the Universe, by way of deduction from perfectly established postulates, such as that one which governs universal gravitation. .*".

Then, after a presentation of the main lines of his approach, he ended his speech as follows:

> "*At the end of last October, some eminent physicists were kind enough to respond to my call and to meet in a "Physics Council", held in Brussels under the chairmanship of Mr. Lorentz from Leiden. I had been informed of a recent result showing the impossibility of reducing black-body radiation laws to the laws of kinetic and electrodynamic theories. I therefore thought that by bringing together the main specialists, I would let them provide some clarifications to these questions.*
>
> *At the same time, the disarray of these formerly prevalent theories, prompted me to abandon a reserve that I had imposed on myself, and to*

publish an overview of the special views I had formed on the subjects of matter, of the ether and of energy (...).

To the intellectual contribution (...) I just outlined, I recently attached a material contribution. Thus, following the recent Physics Council, and with the support of its members, I decided to found an International Institute for Physics, which will be in some way the consecration and extension of the Council, and which will be renewed from time to time. The Institute obviously also targets physical chemistry, and in this sense, I think, Gentlemen, that you will welcome with satisfaction the announcement of this new foundation, intended to help researchers, whatever their nationality and the paths they intend to follow in the pursuit of truth[355]...".

The ISC-members were informed of ISIP's establishment on June 2, 1912. Knudsen, the Committee's secretary (figure 21), sent them a circular letter, written by Lorentz[356]. It stated that the Institute had been founded for a period of thirty years: from May 1, 1912, to April 31, 1941.

FIGURE 21: Martin Knudsen in 1921 (centre of the picture), see figure 41.

The million, paid by Solvay to an Insurance Company, was expected to produce 29 annuities of an amount of 55 332 francs. In addition to this income, ISIP could reckon on an extra annuity for the year 1912-1913.

It had been agreed that one third of the yearly income would be spent in the form of grants. Regarding ISIP's subsidy-program, Lorentz argued that it was imperative to limit the number of applications. He asked the members to determine the areas of research which should be considered for a grant, suggesting that ISIP should favour studies of radiation phenomena (Röntgen rays and radiations emitted by radioactive bodies) and experiments that could be linked to quantum theory or to molecular theories.

The ISC-members were invited to make proposals, and to name Academies and scientific journals which had to be informed of the Institute's birth, and of its objectives (see figure 22).

The Committee's first meeting was planned for the end of September 1912. Its agenda comprised the choice of a subject for the Physics Council of 1913, the setting up of a list of members to invite, and the question of ISIP's first research grants.

One point hadn't been settled: the appointment of the members of the Administrative Commission, a time-consuming operation which required the agreement of the King of the City of Brussels, and of the University.

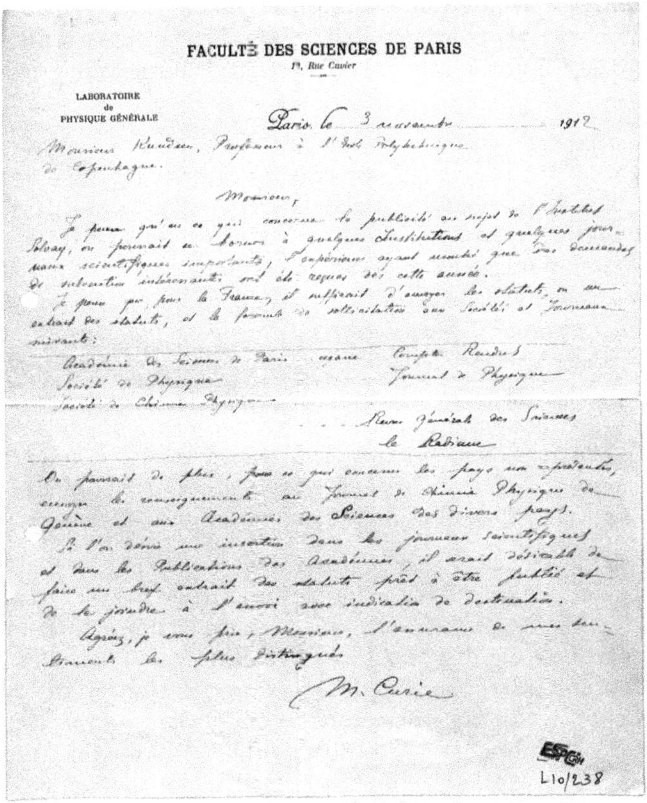

FACULTÉ DES SCIENCES DE PARIS
1ᵗ, Rue Cuvier

LABORATOIRE
de
PHYSIQUE GÉNÉRALE

Paris le 3 novembre 1912

Monsieur Knudsen, Professeur à l'École Polytechnique de Copenhague.

Monsieur,

Je pense qu'en ce qui concerne la publicité au sujet de l'Institut Solvay, on pourrait en donner à quelques Institutions et quelques journaux scientifiques importants, l'expérience ayant montré que des demandes de subvention intéressantes ont été reçues dès cette année.

Je pense que, pour la France, il suffirait d'envoyer les statuts, ou un extrait des statuts, et la formule de sollicitation aux Sociétés et Journaux suivants:

Académie des Sciences de Paris organe Comptes Rendus
Société de Physique Journal de Physique
Société de Chimie Physique
... Revue Générale des Sciences
le Radium

On pourrait de plus, pour ce qui concerne les pays non représentés, envoyer les renseignements au Journal de Chimie Physique de Genève et aux Académies des Sciences des divers pays.

Si l'on donne une insertion dans les journaux scientifiques et dans les Publications des Académies, il serait désirable de faire un bref extrait des statuts prêt à être publié et de le joindre à l'envoi avec indication de destination.

Agréez, je vous prie, Monsieur, l'assurance de mes sentiments les plus distingués

M. Curie

ESPCI
L10/238

FIGURE 22: Letter from Marie Curie to Knudsen of November 3, 1912. Courtesy of ESPCI-PSL, Paris, Centre de ressources historiques (fonds Langevin).

Belgian presence in the ISC

Once more, an unexpected problem caused by Goldschmidt created a last-minute complication. On August 17, Solvay received a letter[357] from Lorentz, informing him of Goldschmidt's great displeasure at having been kept out of the discussions that had led to an operation he had been the first to suggest: the foundation of ISIP.

Lorentz didn't hide his embarrassment. He reminded Solvay that he had been asked to come to an understanding with Héger on the question of ISIP's statutes. Expressing regret for not having consulted the "Council's general manager", he wished to make up for his lack of courtesy towards a colleague who had rendered great services, and suggested[358] to offer him a place in the Administrative Commission.

Unfortunately, the idea didn't suit Solvay, who had foreseen a Commission of three members, two of whom had to be chosen outside his family: one representing the King, the other the University. He therefore proposed[359] an alternative:

> "Regarding your proposal to involve Mr. Goldschmidt one way or another in the work of the Council, I can only agree with you in principle. I have heard of Mr. Goldschmidt's wishes on this subject, and it seems to me to result from a letter he wrote to me that he would like to take care of the material organization of Council meetings as he did for the previous meeting, and for which, according to his expression, he acted as a "general manager". Yet, it seems difficult to me to incorporate him into the Administrative Commission, the organization of which is again on my mind – and which seems to me to be invoked to deal with questions of a very general nature. Maybe it would be better to attach him, in some capacity, to the Scientific Committee itself, so as to enable him to attend your meetings, and to take care of their organization, a matter which concerns directly the Committee and not the Administrative Commission...".

Solvay also mentioned the three Commission members he had in mind. He would ask Tassel, his first adviser for physics, to be his own representative. Professor Jules-Emile Verschaffelt, who was well known in Leiden, seemed a suitable representative of the Université Libre de Bruxelles. Regarding the King's representative, he would make a proposal after having consulted General Jungbluth, the "aide-de-camp" to his Majesty.

In a letter of September 16, Lorentz told[360] Solvay that he sided with his idea about Goldschmidt, and that he had taken the advice of the other Committee members. He was happy to announce that they considered it legitimate to welcome a Belgian representative in the ISC, and that Goldschmidt would do the Committee great service by accepting Solvay's proposal.

Conclusion: Godschmidt's complaints for having been kept out of the realization of his project had a long-term effect: it perpetuated the presence of a Belgian scientist in ISIP's Scientific Committee.

Royal patronage

Reassured by the appropriate solution of the Goldschmidt case, Solvay concentrated on his next task: the constitution of the Administrative Commission. Following his wish to place his new foundation under the King's patronage, he paid a visit to General Jungbluth. He took care to insist on ISIP's international character, and on the presence of a Belgian physicist in its Scientific Committee. In order to convince the King that his high patronage wouldn't set an undesirable precedent, he made clear that the request came from an illustrious foreigner, the Council's chairman H. A. Lorentz, who accepted to lead this Belgian foundation: history's first International Institute for Physics.

Solvay also named Professor Héger as a possible candidate for becoming the King's representative in the board of directors[361], knowing that the idea had a good chance of being approved. On the other hand, he felt confident that the University would accept to be represented by Professor Verschaffelt, a physicist who had worked on a thesis under the direction of Kamerlingh Onnes, and whose candidacy would be supported by Lorentz.

To the rescue of a physics laboratory in Moscow

In his plea for an Institute that would operate during fifty years, Lorentz had insisted on the dramatic situation of the Lebedew laboratory. Deeply moved by his colleague's distress, he would have liked to be able to answer his call immediately. But in March 1912, it wasn't possible: ISIP hadn't yet been founded... Lorentz therefore refrained from mentioning his wish in his letter to Solvay.

In June, things had evolved: ISIP's birth had been officially announced. Lorentz could alert the ISC members to the disastrous situation of the Moscow physicists. In order to do so, he attached to Knudsen's circular letter a copy of Lebedew's appeal, and a shortened version of Ehrenfest's letter of April 24, 1912 (Lorentz had circled his most vehement comments in red[362], and had asked Knudsen to remove them from the text).

Unfortunately, the Moscow situation had seriously worsened, following Lebedew's death in March 1912. In addition, it had become clear that the research carried out in his Moscow laboratory didn't meet the criteria set by the ISC for a subsidy.

However, Lorentz was determined not to give up. He reckoned he could count on Solvay's support, who had written[363] in his letter of May 20: *"I am very sorry to learn of the death of Lebedew, about whom you had told me, and who had previously drawn my attention through his work on the pressure of light. It's very sad indeed..."*.

Aware of the difficulty at hand – Lorentz couldn't suggest cutting ISIP's budget for grants – he proposed to take 8 000 francs from the Institute's operating budget for the year 1912-1913. He told his colleagues that this limited gift would be perceived in Moscow as a mark of esteem, and that the fact that it emanated from an international organization would encourage Muscovite businessmen to step up their support of research at the Lebedew laboratory.

Rutherford (figure 23) was the first to receive Knudsen's circular letter and its attachments. He approved the idea of limiting subsidies to research on specific topics, indicating

FIGURE 23: Ernest Rutherford in 1911 (standing, centre of the picture), see figure 1.

that he would propose work on Röntgen rays, and on beta and gamma radiation emitted by radioactive bodies. Commenting on the situation in Moscow, he deplored the death of Lebedew, a physicist he knew and who enjoyed his confidence. He accepted to provide support to the Moscow team, provided that it be limited to the budget year 1912–1913.

Noteworthy detail: in his letter (figure 24) to Knudsen[364], Rutherford mentioned a recent event:

"*I have one of your countrymen, Dr. Bohr, working with me at the moment. He seems a very able young fellow...*".

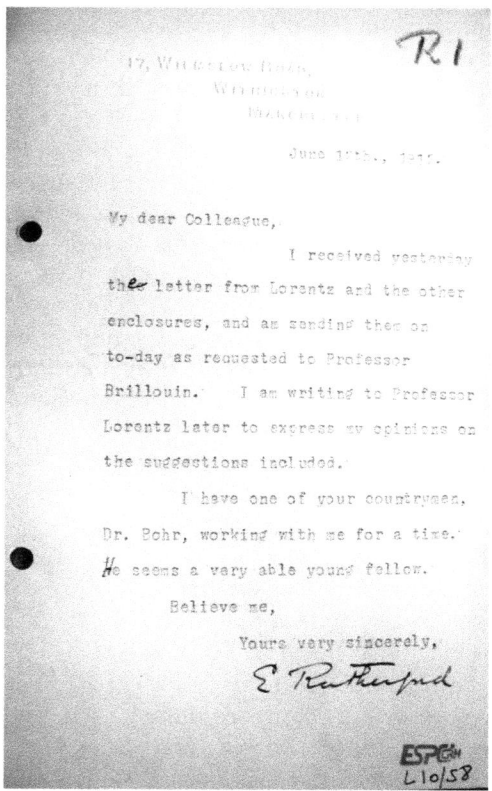

FIGURE 24: Letter from Rutherford to Knudsen of June 12, 1912. Courtesy of ESPCI-PSL, Paris, Centre de ressources historiques (fonds Langevin).

Knudsen's reply (figure 25) to Rutherford is interesting, for it is known that he had no taste for quantum theory, and that he didn't bother to read Bohr's articles[365]:

"*Remember me, please, to my young friend Dr. Bohr when you see him. I am glad to hear that you consider him as an able man...*".

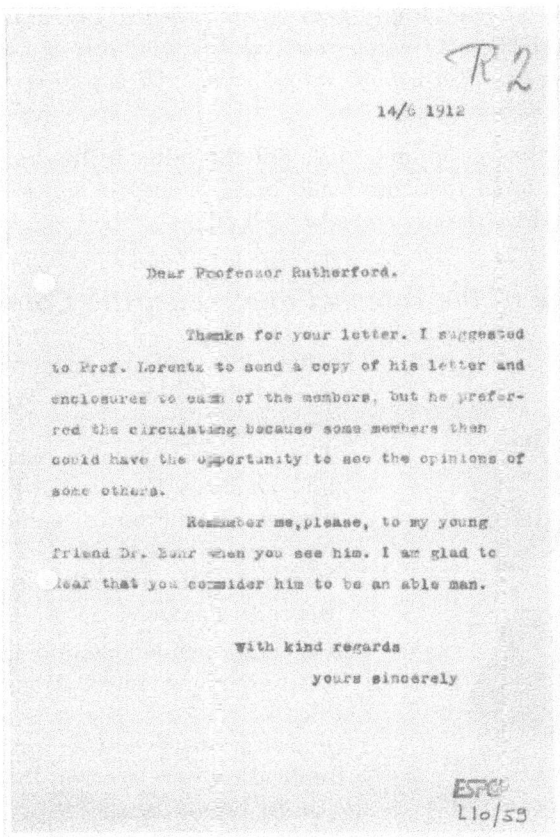

FIGURE 25: Letter from Knudsen to Rutherford of June 14, 1912. Courtesy of ESPCI-PSL, Paris, Centre des ressources historiques (fonds Langevin).

Reacting to the circular letter, Nernst suggested to wait for the first ISC meeting before announcing the Institute's program. Notwithstanding his doubts about the value of the Moscow team, he would follow Lorentz's advice[366].

Building on these positive reactions, Lorentz announced to Solvay that the Committee accepted to provide an exceptional support of 8 000 francs[367] to the Lebedew laboratory.

Solvay marked his approval by telegram[368], and the grant was sent on August 20 to Dr. Lazarew, Lebedew's successor in Moscow.

Thus, on September 16, 1912, Lorentz had the privilege of including these words in his message[369] to Solvay:

"*I still have to tell you that Mr. Lazarew in Moscow acknowledged receipt of 8,000 francs, sent to him by Mr. Knudsen. He expresses the warmest* thanks of the physicists who work at the Lebedew laboratory. The encouragement that their research received has not only a colossal material value, but it has also a huge moral value. All physicists working in the laboratory feel this deeply...".

Indeed, ISIP's first action had a symbolic value: it illustrated the benefits that an International Institute could bring to men of science, whatever the country where their research was carried out.

First meeting of the International Scientific Committee

FIGURE 26: Minutes of ISIP's Scientific Committee meetings: Applications for subsidies. Courtesy of ESPCI-PSL, Paris, Centre de ressources historiques (fonds Langevin).

The first ISC meeting took place two weeks later at Hotel Métropole. Were present on September 30, 1912:
— The Institute's founder: Ernest Solvay.
— ISC members: Lorentz, Brillouin, Warburg, Nernst, Rutherford, Kamerlingh Onnes, Knudsen, Goldschmidt.
— Members of the Administrative Commission: Héger and Tassel.

The agenda included the following points:
1. Statutes of the International Institute for Physics.
2. Organization of the Committees' work.
3. Publication of a Note on ISIP's foundation and on its organization.
4. Allocation of research subsidies.
5. Planning of the next Physics Council: topics to be dealt with, list of members to invite.

Extract from the meeting's minutes[370] (figure 26):

"*Chairman Lorentz opens the meeting and thanks Mr. Solvay for ISIP's creation. He recalls the great loss for science caused by Poincaré's death, and announces the appointment of Mr. Tassel as Mr. Solvay's representative in the Administrative Commission. He also announces the appointment by the King of Mr. Héger as a member of that Commission (the University's representative in the AC hasn't yet been appointed).*

The chairman excuses Mrs. Curie's absence because of health reasons.

The Committee expresses the opinion that the Administrative Commission should be responsible for the election of future ISC members, without the intervention of Academies. However, a decision on this matter is adjourned.

Messrs. Solvay, Héger and Tassel leave the meeting.

Mr. Warburg is elected vice-president of the ISC.

Foundation of the International Solvay Institute for Physics 135

> Mr. Nernst announces that the reports and discussions of the Council of 1911 will be published in German by the Bunsen-Gesellschaft.
>
> The ISC members propose the sending of a Note to inform Academies and Physics Journals of ISIP's creation. This Note, sent by the Administrative Commission should contain the Institute's statutes, the composition of the ISC, and an announcement of ISIP's subsidy program. The applicants should be informed that they must send their requests to President Lorentz before February 1, 1913.
>
> The chairman proposes the holding of a second Physics Council in October 1913, dedicated to questions related to the constitution of matter. This proposal is adopted unanimously."

First research grants

ISIP's subsidy program was the main issue to be discussed during the meeting.

The ISC members decided to limit grants to research on questions that had been addressed during the first Council. They fixed the procedure for the granting of subsidies:
— The President receives the requests and makes the necessary inquiries. These are attached to the requests and sent to the Secretary of the ISC, who sends copies to the Committee members.
— Each ISC member shares his opinion with the Secretary, who transmits it to the other members.
— The members communicate their vote to the President, with an order of priority.
— The President fixes an overall order of priority, on the basis of the individual proposals.
— The ISC proposals are submitted to the Administrative Commission, which adopts the proposed order of priorities, and communicates its decision to the ISC (the subsidies are granted until funds are depleted).
— The ISC Secretary responds to the applicants. The AC Secretary sends the amounts agreed to the selected beneficiaries.

The Committee members examined two subsidy requests which had been sent to Lorentz after Knudsen's circular letter, which encouraged ISC members to make proposals. Both aimed at research on Röntgen rays.

The first request dated from August 17, 1912. It came from Max Laue[371] (figure 27), one of Sommerfeld's young colleagues[372], who

FIGURE 27: Max (von) Laue in 1913 (standing, back of the picture), see figure 30.

had made a major discovery: the diffraction of X-rays by a crystal (it would earn him a Nobel Prize in 1914).

Laue's discovery, the starting point of what is called today "the physics of condensed matter", aroused Einstein's interest, who expressed his admiration in a letter[373] to Ludwig Hopf:

> "*Laue sent me a photograph of his deflection phenomenon with Röntgen-rays. It is the most marvellous thing I have ever seen. Bending at the individual molecules, whose arrangement is thereby revealed. The photograph is sharp, which one hardly would have expected in view of thermal agitation...*".

Laue requested a grant of 4 000 marks (about 6 000 francs) to continue his study. He promised to write a report before November 1913. Lorentz had asked Rutherford to examine the project before the meeting.

FIGURE 28: Charles G. Barkla in 1921 (standing, left hand side of the picture), see figure 41.

The other request emanated from Charles Glover Barkla (figure 28), a professor at London University and an expert on secondary X-rays (he would get a Nobel Prize in 1917). Barkla asked £103, the equivalent of 2 575 francs, for a study of fluorescence phenomena[374].

The Committee suggested, after discussion, to grant 5 000 francs to Laue and 2 500 francs to Barkla.

Conclusion: two ISIP research subsidies were granted before the introduction of the above granting procedure, which guaranteed ISIP's independence from Academies and Faculties (note that the subsidy requests had to be sent to Lorentz's private address, Zijlweg 76, Haarlem).

Launch of "Solvay II"

Following the first Scientific Committee meeting, ISIP's statutes (the final text of which had been established on October 9, 1912) were sent to the ISC members.

The Administrative Commission was constituted three days later. Tassel immediately circulated the Note announcing ISIP's founding and its objectives.

The invitation letters to the second Council[375] were sent on February 12, 1913. Were invited: the first Council members, plus J. J. Thomson, W. C. Röntgen, M. Laue, E. Grüneisen, P. Lenard, W. Voigt, W. H. Bragg, L. Gouy and P. Weiss.

The Council's agenda covered four topics:
1. *Structure of the atom* (report by J. J. Thomson or by Rutherford).
2. *Recently discovered phenomena by Mr. Laue* (report by Laue).
3. *Pyro- and Piezo-electricity* (report by Voigt).
4. *Molecular structure of solid bodies* (report by Einstein or by Grüneisen).

It was agreed that each invitee would receive an indemnity of 750 francs and a copy of the volume of the first Council, published by Gauthier-Villars. The reduction of the indemnity (250 francs less than in 1911) seemed justified by the large number of Council members.

Guest reactions

On April 23, 1913, Tassel announced fifteen positive answers: Bragg, Brillouin, Curie, de Broglie, Gouy, Grüneisen, Hasenöhrl, Knudsen, Laue, Lenard, Lindemann, Rubens, Rutherford, Voigt and Wien. Einstein, Jeans, Kamerlingh Onnes, Sommerfeld, Warburg and Weiss, answered positively a few days later, and so did Goldschmidt, Langevin and Nernst.

J. J. Thomson put a little more time in answering, but finally accepted to present a report. Einstein, on the other hand, declared that he wouldn't report on the molecular structure of solid bodies, nor on any other subject.

Contrary to his first announcement, Lenard made it known that he wouldn't come to Brussels. He later confessed in a letter[376] to Wien, written during WWI, that he didn't enjoy international meetings, and that he couldn't bear the thought of being in the presence of J. J Thomson.

Lorentz took advantage of the presence in Paris of the American physicist Robert W. Wood[377] (figure 29), and sent him an invitation. Wood accepted to come to Brussels. Another last-minute proposal was the programming of a report on the molecular structure of crystals, considered from a chemical point view. It would be presented by William J. Pope from the University of Cambridge, and by William Barlow, a crystallography expert.

FIGURE 29: Robert W. Wood in 1913 (sitting, centre of the picture), see figure 30.

Chapter 5
The second Physics Council

Solvay II took place from October 27 to 31, 1913. The sessions were organized at Park Léopold's Physiology Institute (the building still exists, it houses the Emile Jacqmain School).

5.1 Members and reports

The Council brought together twenty-eight scientists (figure 30). Planck and Röntgen were unable to attend. Voigt announced that he would arrive on October 29.

Eight reports[378] – four in English, three in German and one in French – were presented and discussed:
1. *The Structure of the Atom*, by Sir J. J. Thomson.
2. *Die Interferenzerscheinungen an Röntgenstrahlen, hervorgerufen durch das Raumgitter der Kristalle*, von M. von Laue.
3. *The Reflection of X-rays and the X-ray Spectrometer*, by W. H. Bragg.
4. *The Relation between Crystalline Structure and Chemical Constitution*, by W. Barlow and W. H. Pope.
5. *Quelques considérations sur la structure des cristaux et l'anisotropie des molécules*, par M. Brillouin.
6. *Über die Abhängigkeit der Pyroelektrizität von der Temperatur*, von W. Voigt.
7. *Molekulartheorie der festen Körper*, von E. Grüneisen.
8. *Resonance Radiation and Resonance Spectra*, by R. W. Wood.

Lorentz asked the scientific secretaries, M. de Broglie and F. Lindemann, to take note of all contributions, so as to speed up the publication of the Council's Reports and Discussions.

FIGURE 30: Members of the second Solvay Council on Physics (1913), Brussels, Solvay Physiology Institute, Parc Léopold, S.a.b.ULB. Courtesy of the International Solvay Institutes for Physics and Chemistry.

5.2 Highlights of "Solvay II"

The Rutherford–Thomson confrontation

A major event took place during the discussion of the first report, in which J. J. Thomson (figure 31) mentioned the puzzling deviations that had been observed in the Manchester alpha particle scattering experiments. The author indicated that these findings didn't

FIGURE 31: Joseph J. Thomson in 1913, see figure 30.

constitute proof of the existence of an atomic nucleus, a statement which provoked a lively debate. Rutherford, who had led the Manchester team, refuted Thomson's argument[379] and made a convincing plea in favour of his nucleus hypothesis. It was the first time that the idea was formulated on an international stage (Rutherford had brought the hypothesis up some time before on a British stage, but without a proper discussion of its experimental base[380]).

Solvay II thus became the scene of a historic confrontation between two eminent British physicists. Their debate is a perfect illustration of Solvay's idea: to address major challenges in physics by means of direct discussions between advocates of two opposing views. It also offers us a striking image of "science in progress".

Reports by Bragg and Laue

Another big impression was made by the following reports: Laue's report on the X-ray diffraction by a crystal, and Bragg's (figure 32) on the way in which such X-ray experiments revealed the regular structure of crystals.

As pointed out by Einstein in his letter to Ludwig Hopf, Laue had managed to "kill two birds with one stone". On the one hand he had provided proof that X-rays were a form of light (with a wavelength 5 000 to 10 000 times shorter than visible light). On the other, he had opened the way to a systematic revealing of the atomic distribution in crystals.

W. H. Bragg's advance, together with that of his son W. L. Bragg[381], was brilliantly summarized in their book[382] (published in 1921):

FIGURE 32: William Henry Bragg in 1924 (sitting, centre of the picture), see figure 48.

"*We recently acquired a new method for the study of crystals. Instead of guessing the internal arrangement of the atoms from the crystal's external shape, we can measure the real distances from atom to atom, and draw a diagram representing the structure of the crystal, just as if we were drawing up the plan of a building. The basic idea of this discovery is due to Dr. Laue, professor at the University of Zurich*".

The Nernst–Einstein debate

The earlier disagreement between Einstein and Nernst about the status of the heat theorem took a new turn during the discussion of Grüneisen's report. Nernst led the charge. He said that his point of view had been challenged by Einstein in 1911 on the basis of arguments which, already at the time, seemed doubtful to him...This time, he would refute Einstein's argument by theory as well as by experiment. Thus, basing his proof on a thermodynamic reasoning, he came to the conclusion that *"denying the validity of the heat theorem was equivalent with a rejection of Carnot's principle"*.

Lorentz, who seemed convinced by the demonstration, declared[383]:

"The last proof that Mr. Nernst gave of his theorem, based on the postulate that it must be impossible to reach absolute zero by finite changes, seems to me to be safe from any objection."

But Einstein couldn't agree. He claimed that Nernst's demonstration didn't hold, and that it was preferable to raise the heat theorem to the rank of a postulate. Taking up the chemist's most recent argument, he concluded[384]:

"We thus arrive at formulating Nernst's theorem in a very intuitive way, but unfortunately this again leads to consequences which, by their strange character, arouse suspicion."

Council impressions

Einstein who applauded Laue's discovery, was equally enthusiastic about Bragg's report. He made this clear in a letter[385] to Hopf:

"In Brussels the lecture of Bragg Sr. was extremely interesting. It is unbelievable how much this man has already found out about the lattice structure of crystals and about Röntgen rays. Now, all at once, it is possible to carry out exact determinations of the wavelengths of Röntgen rays. Some metals (e.g., radium) display very narrow emission ranges, so that it is really possible to produce monochromatic Röntgen rays. One can safely say that Laue's figures are now totally explained...".

Einstein also shared his impressions with Elsa in a letter[386] of November 7, 1913:

"It was very interesting in Brussels. I had a heated debate with Nernst, which, oddly enough, did not damage our good relationship. At a banquet, our president H. A. Lorentz deemed it necessary for me to fire off the toast to our host. I did it, but in a quite childish way, because I am singularly inexperienced in such matters. All those who had known me as a quick-witted debater without fear and reproach found it amusing that my mastery of words deserts me so completely when I eat and drink...".

Sommerfeld congratulated himself on the importance of Laue's discovery, made in his Institute. In anticipation of the complement that he intended to bring to Laue's report, he sent a letter[387] to his wife (undated letter, figure 33):

> "I'm still at the hotel, where I will have to leave in ten minutes. I already feel better... Einstein is delicious, as always... we are enchanted by his happy laughs. This afternoon I will take the floor for Laue... My splendid wardrobe will be of no use: there will be no dinner at Solvay's; he invites us at a restaurant...".

Then, in a second letter of October 29th, he told her:

> "The day's work is now over: sessions from 9 a.m. till 1 p.m., lunch until 4 p.m., followed by a road trip to Laeken to visit a wireless telegraphy station[388]. This morning I brought my contribution[389] to the Laue phenomenon: it was good overall, but poorly presented (even if Einstein doesn't think so), because I didn't sleep well... From now on, I will be a listener. It would be better that I give up my planned visit to Debije, and that I spend next Sunday quietly with you (a thought which makes me happy). Voigt and his wife arrived yesterday evening. I still have to write my contribution... The meeting was, of course, of colossal interest.".

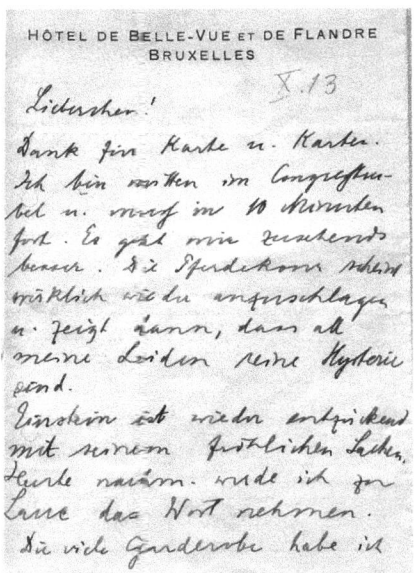

FIGURE 33: Letter from Sommerfeld to his wife, October 1913. Courtesy of Mrs. Monika Baier, grand daughter of Sommerfeld.

Solvay's absence

Contrary to his intentions, Solvay took no part in the second Council. His absence raises questions, for we know from his scientific diary that he spent a lot of time to advance his research in view of this important meeting. We also know that he hadn't given up his plans about a check of his ideas on the source of radioactivity. This is the message[390] he sent to Marie Curie on July 11, 1913:

> *Dear Madame Curie,*
>
> *Nobody wishes more earnestly than I the recovery of your health, from the point of view of resolve that I observed at the Brussels meeting. I therefore hope that the cure and the holidays, announced in your kind letter of the 5th, will all be in favour to you. Hence, it will be in October, or after the Brussels meeting that I will have the great pleasure to visit the installation of your new laboratory, because I intend to stay here until the end of this month. I will be very happy and interested to be allowed to see your laboratory.*
>
> *One word, if you will allow me, about the experiments, so that you can think about it occasionally. I am not at all familiar with the issue of radiation and the use of an electrometric device to observe its variations. I would therefore be happy to know something more about the way in which you understand these things in view of the goal to be reached. My interpretation of radioactive phenomena (I sometimes wonder if I expressed myself clearly) is of a general and extremely simplistic nature. It is a kind of first attempt, such as the one adopted initially by the Curies, if I am well informed. The radioactive bodies are not radioactive by themselves, as is now generally admitted[391]. The bodies draw their radioactive energy from all objects and insofar those already possessed it in their surroundings, which itself earlier obtained it from the sun, direct or indirect...*

Three months later, it wasn't Solvay but one of his collaborators, who contacted Marie Curie with the suggestion of an experiment to be carried out with 1 gram of radium (letter from Herzen[392] of October 15, 1913).

We have no trace of Marie Curie's reaction to these letters, but it is likely that the lack of evidence in support of Solvay's views on radioactivity contributed to his absence at the Council. However, there is a painful event that may have played a decisive role. Ten days after the Brussels meeting, Solvay was cruelly struck by a family tragedy: the death of his grand-daughter, Paule van Parijs (daughter of Jeanne Solvay and of Edouard van Parijs), which occurred at his home on November 10.

We have proof that Solvay was deeply affected by this loss... But could it be that the condition of this young mother – she was 28 eight years old – suddenly worsened towards the end of October, while she was staying with Solvay? If so, this might explain his abatement and diminished interest in the Council's proceedings.

Whatever the true reason for Solvay's absence, one point is clear: the moments of depression in this passionate man were never long.

Thus, in the beginning of 1914, he returned to work and wrote a note on *"masses in self-revolution"*. On April 28, he sent a new letter to Marie Curie, with a proposal which, in his opinion, would confirm the *"ability of radium to transform ambient energy into radioactive energy*[393]*"*. A few weeks later, he returned to the charge, this time with the feeling that *"his role was becoming short"*. These are his words in his letter[394] of May 15, 1914:

> *Dear Madame Curie,*
>
> *I'd be very happy if it were possible for you to help me once more, with Mr. Debierne by experiment and by reasoning on the question* (of the origin of radioactivity), *so as to bring into my mind the complete tranquillity that I seek on this subject. If you could not, for I know your regular occupations and their high value, and if you knew a scientist who would be willing to take charge of this care, I would be extremely grateful if you would kindly inform me. I will assign a bonus of 25,000 francs to obtain full satisfaction on this matter during my lifetime, time and experiments being of course at my expense...*

ISC Meetings

The members of the Scientific Committee met during the Solvay II at Hôtel de Flandre. Their main task was to address ISIP's subsidy program (see chapter 7 for details).

Four meetings took place from October 27 to October 31[395]. They were attended by Lorentz, Marie Curie, Warburg, Brillouin, Nernst, Rutherford, Kamerlingh Onnes, and Knudsen (Langevin, de Broglie, Lindemann, and Verschaffelt only attended the meeting of October 27).

The first point to be discussed was the publication of the Council's Reports and Discussions. Some members noticed that there was no need to translate everything in French (as had been done in 1911 out of courtesy to Solvay); they wished to speed up the publication of the volume. Others defended the idea of a volume in line with the one published in 1912 by Gauthier-Villars. The Committee finally proposed to publish the reports in their original form and to prepare an account of the discussions within the next fifteen days. Yet, the subject remained sensitive. Verschaffelt pointed out that the publication of the Council's Reports and Discussions was a responsibility of the Administrative Commission. He predicted difficulties and suggested not to report on the Committee's decision. Pending a final resolution, Goldschmidt was asked to have a multilingual report printed in Brussels by the publisher Hayez.

We will see in section 8.4 that the publication of this one-of-a-kind volume would be compromised by the Great War and by Belgium's occupation[396].

5.3 Echoes and consequences of Solvay II

A paper by Rutherford on the Council's discussions appeared in *Nature* on November 20, 1913, underlining the interest of Laue's and Bragg's Solvay reports, and praising Pope and Barlow for their analysis of the connection between the structure of crystals and their chemical composition. The author also reported on the events that took place on the sidelines of the Council, such as the dinner offered by Solvay and the visit of a wireless telegraphy station (the one mentioned by Sommerfeld, located in the royal domain of Laeken). This station, noted Rutherford, was able to transmit messages to Congo and to Burma, a real technological feat.

Another paper appeared in *Nature* in May 1914. Signed by Charles Galton Darwin[397], grandson of the famous naturalist, it announced the publication in German of a volume edited by Arnold Eucken[398], which contained the reports and discussions of the first Solvay Council, as well as some developments that intervened between the autumn of 1911 and the summer of 1913. Darwin mentions the Council's results, focusing on issues that had already been discussed in 1911. Concerning the latter, he mentions the works of Debije, Born and von Kàrman on specific heats, Poincaré's proof of the necessity of the quantum hypothesis, the application of the hypothesis to the rotation of molecules in gases... He also expresses one regret: Eucken's volume did not yet contain any reference to Bohr's theory of spectra.

NB: Darwin's criticism could also be applied to the second Solvay Council. The focus of the Thomson–Rutherford debate had been on experimental considerations. No mention of Bohr's theoretical work, which Rutherford had communicated to the *Philosophical Magazine*[399], and which had appeared there in July 1913.

The University Conferences

The smooth running of the first Council organized by ISIP provided proof of the excellence of the "Solvay method". Not less than thirty scientists, including seven Brits, had come to Brussels to attend the meeting... J. J. Thomson had accepted to present a report. The reports had aroused considerable interest, especially from Einstein.

Encouraged by this success, the members of the Administrative Commission decided to strengthen ISIP's visibility in local circles. The idea was to take a cue from the famous "*Friday Evening Discourses*" organized by the Royal Institution of Great Britain since Faraday's time. It materialized in the form of a cycle of "university conferences". Lorentz agreed to start the cycle with two lectures in the spring of 1914. One, intended for a large audience, would take place at the University. The other, targeting an audience of scholars and advanced students, would be organized at Park Léopold's Physiology Institute.

Lorentz chose *Special Relativity* as the subject[400] of his specialized lecture, and *Scientific Prediction*[401] for his large audience conference. The latter took place on March 28, 1914, at Brussels Free University. It was attended by more than hundred personalities: Academy members, professors at Belgian Universities and at the Royal Military School, diplomats and political leaders.

The conference was a great success, but Solvay was absent. He was staying in France to recover from fatigue and from the grief caused by the death of his granddaughter. Tassel, who had been asked to apologize for his absence, sent the following message to Lorentz[402]:

> "*Mr. Solvay, who was extremely tired, and who had been deeply affected by the death of his grand-daughter, left a few days ago for the South of France, where he intends to rest for a few weeks. My colleague Lefébure went to see him, and I just received excellent news about his health...*".

Chapter 6
Foundation of the International Institute for Chemistry

Solvay's fatigue was linked, as we know, to certain difficulties encountered in recent months in his research. But it was caused above all by the countless challenges he had been forced to accept in his attempt to create an International Chemistry Institute. Let us examine the stages of this laborious process.

6.1 Resuming contact with Ostwald

We witnessed Solvay's reaction to Ostwald's call and his promise to devote 250 000 francs to ISIC *"in case of agreement on the foundation conditions"*.

That was in April 1911, before the Physics Council of October–November... Since then, the situation had changed.

Let us go back to the lunch of November 4, and to the discussions which took place between Lorentz and Solvay on the Physics Institute. It is possible that Solvay told his guest about a possible source of inspiration for ISIP: Ostwald's *"Memoir"* on an International Chemistry Institute, received a few days earlier...

One thing is clear: the president of the IACS didn't know of the latest developments. He was far from imagining that his proposal would be eclipsed by a foundation plan for physics. Solvay had no other choice than to open Ostwald's eyes. On November 11, he decided to take the plunge[403]:

> My dear Mr. Ostwald,
>
> I have received your letter of October 29 and the "Memorandum" it announced.
>
> I am now waiting for a second copy. I haven't been able to deal with the matter so far, because I have been mainly absorbed by work related to a Physics Council that I had convened in Brussels. In addition, I had to

> *publish a Gravito-Materialitic study on this occasion. I will take the liberty of sending you a copy.*
>
> *It all took a long time. I have since then been thinking of an international physics foundation which, most probably, would have to be realized at the same time as that relating to chemistry. I will have to examine both things simultaneously, and I won't be able to take sides with one or the other until everything is clarified on both sides.*
>
> *It will therefore be necessary to wait a little if you want my intervention, especially since the time available to me is always limited...*

The message contrasted sharply with that of July 7, 1911. Solvay's enthusiasm had given way to a request for clarification... But the worse was still to come: on January 4, 1912, Lorentz would point out that Ostwald's proposal didn't correspond to Solvay's intentions.

Unaware of the situation, Ostwald convened IACS's first General Assembly. It took place in Berlin on April 13, 1912, bringing together twenty-four representatives of twelve Chemical Societies. The president informed his colleagues of the 250 000 francs that Solvay had accepted to devote to an Institute, destined to become IACS's Central Office. Pleased by the unexpected news – Ostwald didn't specify that the offer was subject to conditions – the delegates thanked the donor in a letter[404] with a photo of the Berlin Assembly.

Solvay immediately sent a note[405] to Lorentz:

"*The gentlemen of the Internationale Association der Chemischen Gesellschaften (Mr. Ostwald former president, Sir William Ramsay actual president) are returning to the offensive for their International Chemistry Institute, and my intention, before going any further, and if you don't see any inconveniences, is to send them the statutes that we have drawn up together, in order to see if we wouldn't create for them a more or less similar foundation...*".

Lorentz agreed. ISIP's statutes were sent to Sir William Ramsay (figure 34) on May 3, 1912, together with a personal comment from Solvay:

"*If the International Institute for Chemistry that you have in mind could be founded on ISIP's model, and if an agreement could be established between the two Institutes, many difficulties to be foreseen could be avoided.*"

FIGURE 34: William Ramsay in 1913, see figure 36.

Ramsay replied that he would give an answer after consulting with the other members of the British board.

Encouraged by this prospect of respite, Solvay returned to his investigations. But his hope was short-lived. In June 1912, he found himself caught up again by Ostwald's project. First through Henri Lafontaine[406], chairman of the International Peace Office, who told him of his concerns about the location of the planned Institute for chemistry:

> "*An article published in a Dutch journal just fell into my hands by the biggest coincidence. It indicates the efforts made by our neighbours to locate in Amsterdam the International Institute for Chemistry Mr. Ostwald told us about...*".

A few weeks later, it was Henri Van Laer, president of Belgium's Chemical Society, who wished to discuss the matter, indicating that the government could make land available for a settlement of the Chemistry Institute in Brussels.

6.2 Intensification of Solvay's work

Solvay granted Van Laer an interview and hastened to return to his research. He started completing his *Gravito-Materialitic* theory, of which only a summary had been provided so far. His efforts to achieve his goal are impressive: more than fifty notes were written between June and December 1912. During a stay in Evian, he addressed the remaining *quantitative problem*, and reflected on the way of deriving a system of units from Kepler's laws.

Solvay's scientific ambitions reached a peak in the summer of 1912. On July 5, he wrote in his daily register:

> "*I leave it to Herzen and to Hostelet to correct themselves what could be wrong in my considerations. I reckon that with the above indications, and with what I provided in Evian, they will be able to execute the overall "Keplero-Materialitic" work I have in mind, so as to make a well-linked whole of it, extremely precise and complete with regard to principles. By doing it well, it seems to me that a summary could be made for after the holidays, and then, in November or December I would be prepared to present it to the French Physical Society, to which – my idea would be – to join the French Chemical Society, and that under the auspices of Henri Poincaré, who would probably accept if I believe the message I got from Mr. Ch. Moureu*[407]. *We would go to Paris for the occasion (...). This would in no way prevent us from preparing for next year's Physics Council of October-November the complete work in detail – and not only the summary of which I spoke before leaving Brussels...*".

Stroke of fate: Poincaré died unexpectedly on July 17, 1912.

The blow was hard. Lorentz expressed his emotion in a letter[408] to Solvay:

"*The news of the completely unforeseen death of our eminent Poincaré has undoubtedly struck you deeply. It's very sad that one of those who took part in our work of last year – and Poincaré's share was of the utmost importance – has already been taken away from us...*".

Solvay replied[409]:

"*I was in Pontresina (Upper Engadine) when Mr. Poincaré died. I had a boundless admiration for him and was extremely moved by this sad event. I sent a telegram to make it known to his family. His daughter, who had come to Brussels on the occasion of the Council, wrote to me to get a special picture from the Council's group, a wish which I am busy realizing...*".

Yet, the industrialist-investigator didn't disarm. Firmly decided to present his theory to the members of the French Physical Society, he summoned his collaborators and specified their tasks. Herzen was in charge of completing the "*theory of gravitational matter*". Hostelet was responsible for extending it, and for drawing from it a proper chemical theory. The idea was to establish a theoretical basis of the *active Universe*. Several notes, written during this period, attest to Solvay's efforts to convince his collaborators of the definitive nature of his results. Faced with their objections he sometimes reacted with irritation, urging them to focus more on his system of interpretative reasonings. But the problem re-emerged. After Tassel's withdrawal, the resistance of younger collaborators remained a source of periodic concern.

6.3 The chemists and their expectations

As Solvay desperately tried to complete his theory, he found himself under new pressure. At the end of August 1912, he got this message from Van Laer[410]:

"*May I remind you the promise you made to send me a copy of ISIP's statutes. I want to actively take care of the Chemistry Institute, according to the views you developed during our last interview...*".

Unwilling to be pushed around, Solvay waited until November 5 before answering. Van Laer received ISIP's statutes (in confidence), together with a personal warning:

"*After having conferred with my colleagues, and contrary to what I initially thought, I would ask you not to rely on these documents in case you intend to act vis-à-vis the International Association of Chemical Societies. I don't want to introduce any personal incitement in this question... The gentlemen from the Committee must have their reasons not to hurry more than they do, and I don't want to disturb them in any way...*".

Two weeks later, Van Laer announced that, in his opinion, an institute "modelled on ISIP" would be much more useful for the progress of chemistry than what had been proposed by Ostwald. Yet, he pointed out that the situation in chemistry was very different from that in physics. In Belgium, chemists were represented by a Chemical Society, capable of formulating a preliminary draft of statutes for an International Institute for Chemistry modelled on ISIP, which could serve as discussion basis in an international meeting.

Reassured by Van Laer's intention to conform to the spirit of his foundation for physics, Solvay replied:

"I don't see any problem, on the contrary, in your getting in touch with Sir William Ramsay on this subject, after having informed me of your plan, and by acting with a certain discretion. You chair the Belgian Chemical Society, and you have previously acted vis-à-vis Mr. Ostwald, the former international president. You are therefore in a good position to act again. The fact that it seems almost sure that these gentlemen from the International Association of Chemical Societies are also working on the question, makes it unlikely by second thought that it upsets your intervention, as long as it occurs in the sense that suits me, according to what you indicated. Probably even, it could be of help to these gentlemen. So, do what you think is best ..".

On December 5, Solvay received Ramsay's long-awaited response.

Cold shower: the new IACS president expressed the wish of the British office that the means made available to the Institute would be used for the constitution of a complete bibliography of all chemical sectors...

Solvay realized that Ramsay's ideas didn't differ much from Ostwald's, and that their goals didn't align with his. He prepared an answer to Ramsay which he submitted to Van Laer. The latter had been busy drawing up a draft statute for ISIC which took the founder's priorities into account, while taking up some ideas put forward by Ostwald. In case of agreement, he declared himself ready to go to London to defend his project.

Van Laer's work was commendable, but his idea of going to London didn't suit Solvay, who wanted to prevent his foundation project from being seen by the English as an initiative of the Belgian Chemical Society... Furthermore, it was clear that the negotiations would be delicate, and that they should be conducted by a man of confidence. Solvay entrusted the mission to Tassel, who examined Van Laer's proposal, and made some changes in his draft statutes. Before leaving for London, he reformulated the letter to Ramsay, specifying Solvay's requirements: regular organization of Chemistry Councils, awarding of research grants on the proposal of an autonomous International Committee, presence in this Committee of a representative of the Belgian Chemical Society. In return, Solvay committed himself to taking into account *"the slight modifications that would have to be made to ISIP's statutes in order to adapt them to the special situation of chemistry"*.

But things didn't go as planned. On December 20, Tassel announced from London that Ramsay was strongly opposed to the organization of regular Councils, and that he didn't believe in the need to encourage research in chemistry by awarding grants.

Fortunately, all wasn't lost. Ramsay proposed to take the advice of Albin Haller, the Association's third man (figure 35).

Solvay, who was in Paris, reacted right away:

> *My dear Tassel,*
>
> *I would be happy if you came to see Haller here. I met him on Saturday evening at Mr. Langlois, the director of the "Revue Générale des Sciences" and I suggested it to him. But first he should take notice of the statutes and regulations of the Physics Institute; please send them to him upon receipt of this letter, and wait until Monday to come to Paris. Write him a note about what Ramsay told you, and offer to go see him on Monday, adding that I am still in Paris and that I could also see him if necessary.*

FIGURE 35: Albin Haller in 1913. Courtesy of ESPCI-PSL, Paris, Centre des ressources historiques.

The following day Tassel got a new letter from Solvay, asking him to bring a copy of the letter sent to Ostwald on June 23, 1911, in which it was specified that the 250 000 francs were to be considered as an "offer subject to conditions".

Tassel prepared his trip to Paris, and addressed this message to Haller:

"Sir William Ramsay formulated several criticisms about the planned organization, and must have written to you... I agreed with him that I would try to see you, in case it would be convenient for you to consider the matter... Maybe we could come up with a project that allows Mr. Solvay to take a decision... Mr. Solvay will stay in Paris until January 15...".

Haller replied[411] on January 2, 1913, inviting Tassel for lunch or for dinner.

6.4 Haller's good offices

Tassel and Albin Haller had been in contact previously. The two men were made for each other.

Of Alsatian origin, Haller had been used to the German system, which favoured a wide range of investigations, from pure research to industrial applications. Before moving to Paris, to take charge of the *Municipal School of Industrial Physics and Chemistry*, he had spent several years in a Lorraine institution, well known to Solvay: the Faculty of Sciences of Nancy. This town was close to Dombasle, location of the first Solvay plant outside Belgium. In 1896, the Nancy Faculty had benefitted from the generosity of the Solvay Company[412] for the establishment of an Electrochemistry Institute. In short, Haller found himself in a favourable position to obtain concessions from Solvay. He was, according to Tassel, *a man who knows to sail.*

This chemist, who would succeed to Ramsay in 1913, was quick to discover Solvay's absolute priority: the establishment in Brussels of a Chemistry Institute, destined to become the Association's Central Office. He decided to grant Solvay satisfaction on this point, in return for his consent to statutes for the Institute that would meet AICS's needs.

Haller's strategy was to operate in several steps:
i. Obtaining Solvay's renouncement to ISIC's subsidy program in exchange for the guarantee that Brussels would become the location of IACS' headquarters.
ii. Isolating Ostwald among IACS-leaders so as to get Haller's draft statutes approved by a majority, including Ramsay and Philippe-Auguste Guye (Haller's colleague from Geneva).
iii. Obtaining payment to the IACS of the annuities produced by the 250 000 francs promised to Ostwald, in exchange of the Association's commitment to hold its annual meetings in Brussels.
iv. Persuading Ostwald to return in view of the majority's wish to entrust ISIC's scientific direction to a "delegation" of the IACS (*i.e.*, by abandoning the idea of an autonomous Scientific Committee).
v. Proposing Brussels for the Association's next General Assembly, in order to obtain Solvay's agreement on measures (i), (iii) and (iv).

Tassel's negotiations with Haller started on January 15, 1913.

Solvay, on the other hand, found himself confronted to a new difficulty. The problem was a letter from Ostwald, who responded to his *"generous plan for chemistry"* by asking a financial contribution to another, more ambitious foundation: *Die Brücke.*

> *"Isn't there a better way to thank you,* wrote Ostwald, *than to ask for something more?... So, if you feel the slightest inclination in favour of the "Bridge", help us; your name will be attached to one of the greatest works of mankind: the organization of the brain of the world* (underlined)".

Slightly more worrying was the fact that the letter contained a postscript:

> *"I will write you soon about Mr. Haller's plan concerning the International Institute for Chemistry. Wouldn't it be good to have a personal meeting in Brussels of Messrs. Ramsay, Haller and myself. I would come with pleasure; it would be an opportunity to see you again..".*

The message was clear: the former president of the IACS tried to regain control.

Solvay was determined not to enter the game. He couldn't accept the idea of seeing himself alone in a discussion with three leaders of the Association... Nor was he willing to embark on a new project. His answer[413] to Ostwald was unambiguous:

"*I can in no way go down the road that you pursue for the "Brücke". My plan of action for the future is clear in my mind, at least roughly: it is limited, and the social organization that you indicate doesn't enter into it... The only thing I can do, to be nice and helpful, is to allocate you 10,000 francs a year, and that for five years, if your organization holds on; we would see later. It would be a way to express my thoughts towards you and towards the work you have undertaken. Regarding the International Chemistry Institute, to be founded in Brussels* (underlined), *and to put me at ease, I have been in touch with your successor, Sir William Ramsay, according to your wishes. Sir William put me in touch with Mr. Haller, and it now is the latter who examines the matter. I only want to deal with it very secondarily, because I am not at that, I already told you often. I have hardly any time available and even that bit is already overloaded. I'll wait for Mr. Haller's proposals. I need something simple and constant, like for the Physics Institute, otherwise I can't consider it. Think carefully about my special situation, my dear Ostwald, and you will understand me. I must hold on, I'm getting old...*".

Tassel, for his part, hastened[414] to inform Haller of the latest developments:

"*I believe I'm doing well by sending you, in confidence, a copy of the letter that Mr. Solvay just received from Mr. Ostwald, as well as a copy of Mr. Solvay's reply, for they both deal with the International Chemistry Institute, to the realization of which you are willing to collaborate. The documents attached to Mr. Ostwald's letter aimed at obtaining of a subsidy of 1 million for the Brücke...*".

On January 24, Ostwald sent a new letter[415] to Solvay, criticizing Haller's plan:

"*In Mr. Haller's project the Institute would be, in a way, a copy of the Carnegie Institute, which has been operating for five or seven years with huge resources to help individual researchers... My thought would be to create something totally new, that is the systematic organization of a whole science. I shall write to Haller when I shall find the time and the strength...*".

Solvay reacted the next day with a hint of irritation[416]:

"*It is me who wanted the planned Institute to be modelled on that of physics, and this for the reasons I gave you in my previous letter. Even I want this question to end without much delay, one way or the other, you will understand this, always for the same reasons...*".

As to Haller, he continued on his chosen path, trying to get Ramsay's and Guye's support for his plan. On January 28, he informed Tassel of his early successes:

> "I'm still waiting for Ostwald's reply...I already got the opinion of Ramsay and his colleagues, who accept the whole project".

On February 6, Tassel reported to Solvay from Paris[417]:

> "I managed this morning to see Mr. Haller... Messrs. Ramsay and Guye slightly redrafted his project. As for Mr. Ostwald, he didn't give any sign of life. Anyway, the matter seems to be on the right track: Mr. Haller seems to have sailed perfectly to obtain Mr. Ramsay's support and that of the English group...".

6.5 Solvay's personal research

For Solvay each period of respite was an opportunity to reactivate his own research. In January 1913, he thought of applying his *Gravito-Materialitic* theory to the rotation and the orbital revolution of planets. His work on the "*Universal Cycle*" testifies to the influence[418] of Claude-Louis Berthollet, a leading French chemist of the late eighteenth century who studied the natural formation of soda[419] in Egypt.

According to Berthollet, the mechanisms that govern the cosmos must be similar to the ones which govern molecular processes. This belief is clearly expressed in his publication "Chemical Equilibria":

> "The forces which produce chemical phenomena are all derived from the mutual attraction of molecules in bodies, to which the name "affinity" has been given to distinguish it from astronomical attraction... However, since it is very probable that the affinity doesn't differ in its origin from general attraction, it must also be subject to the laws which mechanics has determined for the phenomena due to the attraction of masses, and it is natural to think that the more general the principles achieved by chemical theory will be, the more analogies they will show with mechanics".

One of Solvay's goals in molecular theory was to grasp the meaning of absolute zero, and to provide a *Gravito-Materialitic* basis to the van der Waals' law of corresponding states[420]. His wish to have his views confirmed by suggested experiments, was triggered by James Dewar's paper "*New Phenomena Observed in the vicinity of the Absolute Zero*", a translation of which had appeared in the "*Moniteur scientifique du Docteur Quesneville*[421]". Solvay was prepared to call on the expertise of members of ISIP's Scientific Committee:

> "Dewar's article from the "*Moniteur scientifique*", in which the question of the critical point of simple bodies is thoroughly discussed, leads me

to restate my point of view in this regard (...). We could say that all simple bodies are, at their critical point, in their corresponding state, and therefore quite comparable to each other as to their physical and chemical properties (...). The experimental observations that I suggest here (together with the others already envisaged) would serve to assert more and more my Gravito-Materialitic theory, and I wonder if I should point them out to Mr. Lorentz, who could perhaps designate an experimenter to carry them out.

Mr. Nernst would probably do it well[422]. The considerations of Dewar's article show strongly that a re-examination of fundamental physics is more and more necessary; it is the task to which I have been attached for thirty years, specifically foreseeing the situation that presents itself...".

6.6 Back to ISIC

Haller, as we saw, "*sailed perfectly to obtain Ramsay's support to his draft statutes for ISIC*". However, his position with respect to the planned Institute didn't differ much from that of his colleagues. This clearly emerges from Haller's letter to Ostwald of March 3, 1913:

My dear Colleague,

The draft statutes that I had the honour to communicate to you were submitted to Messrs. Ramsay and Guye who, apart from a few formal observations which didn't touch on the very substance of the project, gave their consent. Like me, both feel that from the very point of view of our Association of Chemical Societies, it is a piece of good luck that Mr. Solvay intends to create an International Institute for Chemistry, because our Association will be able to benefit, to some extent, from funds that will be devoted to the administrative part of this Institute. Without doubt, it would be better for us that the sum of 1 million which he wants to allocate to this Institute, with all the modalities which appear in the statutes of the Physics Institute, were allotted to us without other general conditions than those that govern our Association. I agree with you, in particular, that the subsidies granted for a few thousand francs to researchers often don't produce results in proportion to the sacrifices made.

On the other hand, like you, I believe that the aim of our Association is not to instigate research, nor to give grants to young people wishing to improve their skills in foreign laboratories. But we shouldn't forget that the IACS and the International Chemistry Institute are different institutions by their origin and by their purpose. The first, formed as a result of our joint initiative, is a community full of good will and generous intentions, but which has neither funds nor premises. The second will be an emanation of a man with broad ideas, of a great benefactor who wants to put part of his immense fortune at the service of science and of the youth destined to cultivate this science. Is it not legitimate for Mr. Solvay to put conditions on his donation, and to see that his nationals would benefit from it to a small extent?

It was by drawing inspiration from his wishes, as well as from the statutes of the International Institute for Physics, that I wrote my project.

Now, we are quite prepared, Mr. Ramsay and I, to plan with you a visit to Mr. Solvay so as to make him reconsider his decision, and to find the means of convincing him to grant us the million without any other condition than that it be devoted to the well of chemical science. But to get there you would have to notify the generous patron of our visit, and to make him understand the goal that we are pursuing, without deviating from this goal. Indeed, I do not believe that it is useful or necessary to think of a large Institute provided with laboratories and collection rooms intended for research. A place with a meeting room, library and staff to collect and classify the documents that we will use for our work, would suffice. This place will be the seat of our secretariat and may, even should, be located in Brussels. We may well make this concession to Mr. Solvay if he enters into our views. In the event that he maintains his own, we would adapt to them by suggesting statutes of the kind I submitted to you.

Please think about all these questions and give us an answer so that we can make our arrangements in the event of a trip to Brussels.

That the two founders of the IACS shared the same views on the needs of chemistry shouldn't come as a surprise. Their priorities reflected the expectations of a large community of chemists, which differed considerably from those of the much smaller community of physicists. This difference between the world of chemistry and that of physics was bound to hinder Solvay's wish to create two international Institutes on the same model.

In addition to this basic difficulty, there was the indisputable fact that Haller was a better negotiator than Ostwald. Thanks to his trustful relationship with Tassel, the Alsatian quickly realized that there were limits that Solvay wouldn't cross.

In March 1913, his strategy began to pay off. A majority within the IACS seemed ready to accept a compromise. It provided for an allocation of two-thirds of ISIC's annual income to subsidies and to scholarships. On the other hand, the remaining one-third for IACS would be increased with the interest produced by the promised 250 000 francs. Should Solvay agree with this last point, the Association would consider setting up its headquarters in Brussels.

On March 13, Haller announced to Tassel that Ramsay and Guye had communicated their final remarks on his draft statutes. He also mentioned a letter from Ostwald.

One month later, Haller sent[423] a proposal to Brussels, together with a copy of his letter to Ostwald and a translation of the latter's response. Taking up Ostwald's idea of a meeting with Solvay, he declared:

> "In the event that none of the proposed solutions satisfies Mr. Solvay, we are quite willing to go to Brussels."

Tassel hastened to send a telegram to Solvay:

> "Received project Haller and observations Ostwald; must I come to Paris or wait in Brussels for your return?"

The answer was: "*Wait for my return*".

An agreement was finally reached at the end of April 1913. Tassel announced to Haller that his draft statutes for ISIC had been approved by Solvay. They provided for the establishment in Brussels of the Association's Central Office for a period of thirty years (ISIC's lifespan) and for a seat for a Belgian representative in ISIC's Scientific Committee. They also specified that two thirds of the Institute's annual income would be made available to this Committee.

However, there was a price to pay: the statutes didn't provide for research grants.

6.7 Impact of Ostwald's observations

Solvay's approval of the deal raises an obvious question: why did ISIC's founder waive one of his main objectives?

An answer may be found in Ostwald's reaction to Haller's letter of March 3, 1913 (a plea in favour of his own proposal, transmitted to Tassel):

> "*To the many grandiose foundations that Mr. Ernest Solvay has established for the benefit of science, by means of his personal fortune, he has added a new one for 1 million francs, and intended for an International Institute for Chemistry.*
>
> *The trusted persons he consulted on the use to be given to this sum made the commitment to shed light, as consciously as possible, on all the aspects of the problem in order to guarantee a maximal useful effect of the immense tool that has been placed in their hands, and to transform the enormous energy*[424] *represented by 1 million francs into maximal useful work for the benefit of chemical science.*
>
> *The first idea which comes to mind, and which Professor Haller has expressed in an absolutely rational way by his draft statutes, is based on the hypothesis that for the development of chemistry it would be better to financially support young researchers who have special aptitudes in this field, in analogy with what has been achieved by the statutes of the Sister Foundation for physics.*

However, the two sciences – physics and chemistry – are very different from the point of view of their culture. While the number of those who study physics isn't large, and the possibility of finding an opportunity for the physicist to continue his research is by no means wide, the situation in chemistry is, in this respect, completely different.

As a result of the enormous development of the chemical industry, owing to the application of chemistry to all possible problems of economic and social life, not only is the number of purely scientific chemists much larger, even out of all proportion, than that of physicists, but the opportunities created are also much more numerous, also out of all proportion, in the form of laboratories and institutes where they can pursue their scientific work...

There are everywhere the most abundant and varied opportunities for putting into practice the knowledge of chemistry and for engaging in research in chemistry, so that a further contribution to increase this possibility will only bring about very little change in the existing situation, and doesn't in all cases provide the equivalent of a fundamental and essential support for science .

The Academies in Paris and Berlin have, for example, very considerable sums to support scientific pursuits of all kinds. And these sums are still far exceeded by various foundations with the same purpose, among which the Carnegie Institute can be considered as the richest and the most active...

Even if Mr. Solvay's endowment were to be used for such purposes, it would only be one institution among many... On the other hand, precisely for chemistry as a science, there are other very urgent needs that the International Association of Chemical Societies, founded two years ago, has intended to meet... The need, according to the Associations' general provisions, to appoint every year a new president, from one or another country, the lack of capital, of a separate building. The latter's absence will permanently prevent library and archives belonging to the Association itself from providing any notable service, or at least will only allow it at the cost of considerable effort.

But if the sum made available by Mr. Solvay were used to create in Brussels a Central Office for the IACS, which represents the totality of chemists working scientifically on the Earth's surface, and if the means were employed to maintain a regular body of collaborators, suitable for the solution of the various problems in this Central Office, this would advance their solution firmly, and chemistry, as a science, would acquire in a short time a situation that no other science possess from the organizational point of view.

At the same time, the name Solvay would be attached in perpetuity to these absolutely new institutions which establish a new era in the scientific work of mankind, a distinction which the donor would deserve all the more since in all his foundations so far, he allowed himself to be guided by general ideas of an extraordinary scope, considerably exceeding the intellectual work of his contemporaries."

There is little doubt that Ostwald's words impressed Solvay and caused him to reconsider his position. One thing is clear: his approval of Haller's draft statutes opened the way to a series of decisive events:
- May 8: Ostwald announced his acceptance of Haller's final proposal.
- May 15: Solvay to Ostwald: "*Your observations seemed to me well founded, and I did what I could in order to endorse them...*".
- May 17: Tassel to Haller: "*Mr. Solvay received a letter yesterday from Mr. Ostwald who seems delighted, even somewhat enthusiastic about the new draft statutes that you communicated to him*".
- July 16: Solvay announced the transfer of 1 million francs to an insurance company.

Conclusion: An agreement was reached on May 8, 1913. It provided that upon acceptance of Haller's draft statutes by the IACS, Solvay would endow ISIC with 1 million francs (to ensure the payment of twenty-eight annuities from May 1, 1914), and that he would make 250 000 francs available to the Association, with the understanding that the income of this sum would be used to cover general expenses.

Haller's wishes had been fulfilled: ISIC's statutes didn't provide for the granting of subsidies, nor for the organization of chemistry Councils. They didn't even mention the setting up of an autonomous Committee, responsible for the Institute's scientific management... This unilateral "adaptation" of what had been agreed with Tassel would only be announced ten days later.

Guye, who had been kept informed of the latest developments, was struck by the importance of the concessions made by Solvay. This is the statement which he made in a letter[425] to Haller of May 7, 1913:

"*My warmest congratulations on the great success of your negotiations with Mr. Solvay. I never thought you would obtain so much...*".

6.8 Haller's master card

On May 18, Haller revealed to Tassel the last-minute change in ISIC's draft statutes which gave him Ostwald's consent. He explained that the agreed proposal had received Ramsay's and Guye's full approval after the replacement of the Institute's Scientific Committee by a *delegation* of the IACS.

Haller minimized the importance of the change, which was nothing more than a simplification, *one cog less*, since it had been agreed that the

Committee members would be appointed on the Association's proposal... Then, playing his last card, he added:

> "We all are therefore in agreement and can definitively draw up our project, so as to be able to present it to the next meeting of our Association, which I propose to have in Brussels next September, if Mr. Solvay sees no inconvenience...".

The manoeuvre was a masterstroke. By choosing Brussels, instead of London, for the Association's next General Assembly, this Council, scheduled for September 1913, would coincide with the celebration of Solvay's golden wedding anniversary, and with the 50th anniversary of his company.

On May 25, Haller went to see the industrialist, who received him for dinner at the château de la Hulpe. The two men agreed on the date of the IACS Council: it would be held from September 19 to 23, Solvay's jubilee celebration being scheduled for September 20.

Back in Paris, Haller asked Tassel to have 100 copies of ISIC's draft statutes printed. He also announced his wish to trigger expressions of sympathy on the occasion of the founder's jubilee.

Solvay, on his part, asked the King to kindly grant his patronage to the new Institute. On August 19, 1913, he received the royal response:

> "I am very flattered that you associate me with the functioning of these important scientific foundations, by which you give a magnificent example of the noblest use of wealth.'

6.9 Culmination of a jubilee

The delegates of thirteen Chemical Societies (figure 36) met in Brussels on September 19, 1913. They represented Austria, Belgium, England, France, Germany, Italy, Japan, the Netherlands, Norway, Russia, Spain, Switzerland, and the United States.

The Council sessions took place at the Solvay Physiology Institute. The members were asked to approve ISIC's draft statutes and to appoint Haller as the Association's president for the year 1913-1914. Several questions were addressed: standardization of chemical symbols, revision of chemical nomenclature, and adoption of the statutes of the International Atomic Weights Committee.

The presence of eminent chemists enhanced Solvay's golden anniversary and his industrial jubilee. The Council brilliantly underlined ISIC's foundation.

A party was held in the Company's offices on September 20, 1913. A second royal telegram accompanied Solvay's nomination as Grand Officer of the Order of Leopold.

FIGURE 36: Members of the second Council of the IACS, Brussels, Solvay Physiology Institute, Parc Léopold. Archives of the Belgian Chemical Society, S.a.b.ULB.

Sir William Ramsay paid tribute to the generous patron of science on behalf of the International Association of Chemical Societies. Other special tributes were paid by the German Chemical Society, the University of Paris, the University of Nancy, and the "Société d'Encouragement pour l'Industrie Nationale". Haller had been asked to present two awards to Solvay: the Golden Medal of the French Chemical Society, bearing Nicolas Leblanc's effigy, and the "Grande Médaille Lavoisier", on behalf of the Paris Academy of Sciences[426]. Lorentz congratulated the industrialist, as chairman of ISIP's Scientific Committee.

The Council members were invited on September 21 to a reception at Solvay's residence, rue des Champs Élysées.

A paper entitled "*L'institut international de physique. L'organisation du travail scientifique et les Instituts Solvay*" was published on October 25, by the French journal *La Nature*. It reported the distinctions granted to Solvay on the occasion of the fiftieth anniversary of the foundation of his Company.

The paper didn't mention the 1 250 000 francs intended for ISIC, but announced the distribution by Solvay of a sum of 5 million francs in donations. These included 500 000 francs to the University of Paris for the development of an Institute of Applied Chemistry, 500 000 francs to the University of Nancy for its Electrotechnical Institute; 500 000 francs to the International Hygiene Congress for its quadrennial prizes; 250 000 francs to the Charleroi University of Labor, 1 million for Workers' Education Charities, and 100 000 francs to the Belgian League against Tuberculosis.

Concerns after the tributes

The tributes of September 20 didn't dispel Solvay's concerns about the resolutions adopted by the Association's Council. Three days after the party he asked Tassel to find out what had been decided about the location of the future IACS assemblies. These meetings, in his eyes, could only take place in Brussels, location of IACS's Central Office.

Contacted by Tassel, Haller explained that he defended Solvay's point of view, but that some Council members disagreed... He therefore proposed a rule that the Association's annual meetings would be held in Brussels, at least once in two. This provision, approved by the Council, would be recorded as follows in the minutes:

> "The meetings of the Association will take place, as far as possible, at least once in two in Brussels."

Haller declared that this was a *"minimum to which he would hold"*.

Other issues, such as the legal status of the IACS hadn't been decided. They would be re-examined on the occasion of a revision of the Association's statutes, scheduled for 1914. Also, financial questions remained to be regulated: the payment to the IACS of the income of the 250 000 francs, and ISIC's provision of the annual income of the million deposited by Solvay.

Haller wished to transform the Association into a Belgian institution, domiciled in Brussels, but this idea came up against a major difficulty: Belgian legislation didn't allow civil personification to be granted to an association. The new president would therefore take steps to provide the IACS with a legal existence in France.

Role and prudence of Tassel

ISIC's financial situation was discussed by the Administrative Commission in March 1914. Tassel, the Commission's secretary, was asked to prepare two letters. One, intended for Haller, would be signed by Solvay. It was a reminder of the commitments made by the Association: the establishment of its Central Office in Brussels, and the holding of its annual meetings at Park Leopold, at least once in two. The author stated that these conditions were to be seen as a *"counterpart to the sacrifices made"*. The letter also indicated that no payment would be made before having received a formal guarantee from the Association that it would respect its commitments (a measure justified by the rotating nature of the presidency).

Tassel's second letter was a note from Solvay to his representative, specifying that payments would be subject to compliance with the above commitments.

More important was the fact that Solvay's representative foresaw a possible collapse of the Association during the Institute's lifetime. He therefore added a clause to ISIC's statutes, which stated that in the event of a dissolution of the IACS the Administrative Commission would entrust the Institute's management to an "International Scientific Committee" of the kind that governed ISIP.

Solvay's letter was sent to Haller on March 11, 1914. The latter replied that he would add to the Association's statutes the resolution[427] that had been adopted during the Council of September 1913: "*The Association will hold its meetings in Brussels,* **as much as possible** *every other time*".

Pretending to ignore the subtle step back from his previous statements, Haller added:

"*This provision is already included in the statutes of the Solvay Chemistry Institute, and I will make sure, as far as I am concerned, that there is harmony between the two texts.*"

Wishful thinking... Haller's promises had no effect. The revision of the Association's statutes, made necessary by Solvay's donation, didn't take place. The General Assembly scheduled for September 1914 was cancelled due to the outbreak of the First World War.

Chapter 7
The Solvay subsidies

The Great War had a lasting impact on the activity of Solvay's International Institutes. It marked the end of ISIP's subsidy program, inaugurated in 1912 to support research projects from all over the world. This program was remarkable by its novelty and by the unprecedented rigor of the allocation procedure.

Curiously, this major task of ISIP's International Scientific Committee (ISC), and the subsidies that were allocated, have never been the subject of a substantial report. We therefore considered it our task to fill this gap in ISIP's story.

For the sake of clarity, we will start with a global presentation of the Solvay subsidies paid during three budgetary years (1912–1913, 1913–1914, 1914–1915). We will then examine the procedure developed by Lorentz to set the order of priorities.

7.1 Global situation

Let us recall the decision taken by the ISC during its first meeting, on September 30, 1912, to grant two research subsidies: one to Max (von) Laue, the other to Charles G. Barkla.

Shortly before, the ISC members had accepted Lorentz's proposal to provide financial support to Dr. Lazarew, head of the Lebedew team in Moscow.

These decisions were taken before the announcement that subsidy requests could be submitted to the ISC chairman. Thanks to the publication of an official note on ISIP's subsidy program, candidates were informed during the autumn of 1912 that they could apply for a grant on the basis of a research project.

The Scientific Committee would accept projects linked to issues discussed at the first Solvay Council. Applicants were invited to send their requests to chairman Lorentz's private address before February 1, 1913.

ISC evaluation of the first requests took place by correspondence. Authors of successful projects were informed by Knudsen that a subsidy would be granted in August 1913. Later evaluations were made in Brussels during Committee meetings. Four such meetings were organized before the First World War[428].

Over the years a total of 97 requests were taken into consideration (some had been immediately discarded by Lorentz). They gave rise to 40 subsidies (52 requests were rejected; 2 were sent back with suggestions for improvement; 3 were postponed or put on hold).

It was agreed that Knudsen would respond to all applicants. Successful ones would be informed of the amount granted (cheques would be sent to them by the Administrative Commission).

List of beneficiaries of a Solvay subsidy (their projects are indicated, as well as the amounts in Belgian francs).

First budget year (1912–1913)

Subsidies granted in 1912, before the application of the officially published procedure (they are indicated with the letters a, b and c):
  a) Lebedew, *Relationship between matter and Hertz waves;* 8 000 (Russia).
  b) Laue, *Study of the diffraction of Röntgen rays by crystals* (a phenomenon discovered by the applicant); 5 000 (Switzerland).
  c) Barkla, *Study of certain fluorescence phenomena;* 2 500 (United Kingdom).

Subsidies granted in 1913, in order of priority (order established by Lorentz on the basis of individual evaluations supplied to him by the ISC members).
  a) Sommerfeld, *Continuation of Mr. Friedrich's research on Röntgen-ray interference and diffraction;* 4 000 (Germany).
  b) Chéneveau, *Determination of the refractive power of salts in a solid state and in a dissolved state;* 1 500 (France).
  c) Hupka and Müller, *Construction of a very high voltage electric machine which will be used for research on cathode rays and other rays;* 4 000 (Germany).
  d) Von Dechend and Hammer, *Research on canal rays;* 1 000 (Germany).
  e) Icole, *Research on thermal conductivity;* 450 (France).
  f) Julius Meyer, *Research on the thermodynamic properties of gases;* 1 600 (Germany).
  g) Edgar Meyer, *Research on the nature of gamma rays;* 3 000 (Germany).
  h) Dember, *Research on canal rays and on Röntgen rays produced in the photoelectric effect;* 1 500 (Germany).
  i) Millochau, *Research on electric discharge;* 1500 (France).

j) Franck and Hertz, *Research on the ionization energy and the free path of electrons in gases*; 2 000 (Germany).
k) Freundlich, *Research on the chemical nature of radioactive bodies*; 1 250 (Germany).
l) G. C. Schmidt, *Research on the fluorescence of sulfur and selenium*; 750 (Germany).
m) Gockel, *Research on the penetrating radiation that exists in the atmosphere*; 800 (Germany).
n) Jorissen, *Research on the chemical action of radium rays*; 500 (the Netherlands).
o) Lowry, *Research on the rotation of the polarization plane of ultraviolet rays in quartz*; 1 000 (United Kingdom).

Total amount: 40 350 francs for 18 successful applications (10 from Germany, 3 from France, 2 from the United Kingdom, 1 from Switzerland, 1 from the Netherlands, 1 from Russia).

Second budget year (1913–1914)

Projects selected for a subsidy in October 1913 (amounts paid in November 1913):

a) Koenigsberger, *Research on canal rays*; 1 000 (Germany).
b) Salmon, *Study of reactions in an electric arc*; 500 (France).
c) Beatty, *Research on cathode rays and Röntgen rays*; 500 (United Kingdom).
d) Bestelmeyer, *Determination of the electric charge of the electron (cathode rays and radiations emitted by radioactive bodies)*; 3 750 (Germany).
e) Dunoyer, *Study of optic resonance*; 2 700 (France).
f) Fajans, *Research on filiation relationships between elements*; 500 (Germany).
g) Fournier d'Albe, *Research on radiation: quantitative study of the absorption lines in the ultraviolet and infrared domains*; 1 200 (United Kingdom).
h) Grummach, *Research on capillarity, on electric resistance and on its dependence of magnetic forces*; 1 000 (Germany).
i) Hausser, *Research on phosphorescence*; 2 300 (Germany).
j) Lowry, *Research on anomalous rotatory dispersion*; 500 (United Kingdom).
k) Moseley, *Research on Röntgen ray deflection*; 1 000 (United Kingdom).
l) Rohn, *Determination of the energy distribution among the spectral rays of the same series*; 1 300 (Germany).
m) Trautz, *Research on the speed of reactions*; 1 500 (Germany).

Subsidies granted in February 1914 (on proposal of the ISC):
a) Laue, *Continuation of his research on the dispersion of Röntgen rays*; 1 000 (Switzerland).
b) W. L. Bragg, *Research on Röntgen rays*; 1 250 (United Kingdom).
c) Stark, *Research on the splitting of spectral lines in an electric field*; 1 500 (Germany).

Total amount: 21 500 francs for 16 requests (8 from Germany; 5 from the United Kingdom; 2 from France; 1 from Switzerland).

Third budget year (1914-1915)

May 1914: Decision to grant an additional subsidy of 4 500 francs to Stark for the continuation of his research.

June 1914: Decision to spend 13 500 francs (part of the remaining 20 500 francs) as follows:

a) Danysz and Wertenstein, *Research on radioactive recoil rays and on beta radiation of the radium family*; 4 000 (Poland).
b) Wien, *Research on positive rays and on the spectral analysis of polarized Röntgen rays*; 4 000 (Germany).
c) Wood, *Continuation of his research on radiation*; 3 000 (United States).
d) R. Seeliger, *Experiments on the luminosity of gases under the action of cathode rays*; 1 000 (Germany).
e) Zemplén, *Research on the longitudinal mass of the electron*; 1 500 (Hungary).

Three requests remained pending. The ISC decided to address the authors and/or third parties for further information. Result: 7 000 francs remained in reserve.

Total amount (assuming that the indicated amounts were paid): 18 000 francs for 6 requests (3 from Germany; 1 from Poland; 1 from the United States; 1 from Hungary).

Global result of three budget years: 40 subsidies for a total amount of 79 850 francs.

One quarter of the selected projects aimed at research on radiation phenomena (Röntgen rays and other rays).

Distribution among countries: Germany 21, United Kingdom 7, France 5, Switzerland 2, the Netherlands 1, Poland 1, Hungary 1, United States 1, Russia 1.

A fruitful surge of generosity

The announcement of ISIP's subsidy program had effects of unforeseen magnitude. On January 29, 1913, Lorentz acknowledged receipt of twenty-five requests, most of which seemed quite interesting. It meant that the Scientific Committee would have to make a careful selection.

The Solvay subsidies 171

On March 2, Brillouin announced in a letter to Solvay[429] that he examined seven of the forty-three requests registered by the ISC, and that none of the proposed projects were to be dismissed.

Solvay replied[430]:

> "It is of course with the greatest pleasure that I notice that the new International Institute for Physics attracts the attention of researchers. Perhaps, to put the Institute in a good position this first year, I could think of helping the Scientific Committee. I shall see what I can do with the Administrative Commission...".

Tassel, who had been kept informed by Lorentz, hastened to announce him[431] the great news:

> "I told Mr. Solvay what you said in your letter of January 29, 1913, about the large number of interesting requests for subsidies which you received as a result of the notice that was sent by the Institute. Mr. Brillouin also said a few words about it in a letter which he addressed to him a few days ago.
>
> Mr. Solvay, who already kindly promised to intervene this year in the expenses of the "Physics Council" in the event – certain moreover – that our resources would be insufficient to fully cover the expenses, seems also willing to intervene in a manner that allows the Scientific Committee to go this year a little further in granting subsidies than permitted by the available resources. Of course, it is only on a purely exceptional basis that Mr. Solvay would intervene, desiring to make matters as easy as possible for the Institute's first operational year.
>
> I think that when the Scientific Committee will have examined the submitted requests, and separated those judged most worthy of support, you could effectively group in two categories the subsidies that you would consider appropriate to see granted. One of these categories would comprise subsidies to be taken anyway from the 17 500 francs available for this purpose, the other would contain subsidies that you would have granted this year if the resources were more extensive. I could then submit these proposals to Mr. Solvay, who would decide to what extent he could intervene before the meeting of the Administrative Commission. The Commission would allocate the subsidies accordingly...".

7.2 How to define the order of priorities?

The increase in the subsidy budget for ISIP's first year[432] was certainly good news. But it didn't reduce the task of the Scientific Committee, responsible for the evaluation of the research projects and for their ranking in order of priority.

A procedure was established on Lorentz's initiative. Each request would be examined by a Committee member, who would write a report. The reports would be submitted to the other members, who would make comments, orally or by letter. A list of priorities would be established by the chairman on the basis of individual priority proposals. This would usually be done on the occasion of a meeting in Brussels.

Ranking of requests

Of the 44 registered applications, 22 were immediately rejected on the basis of poor quality or inappropriate content. For the 22 remaining requests, indicated in the table (figure 37), with the applicant's names, an order of priority had to be established. The available budget would be spent according to this order. After exhaustion of resources, funding of the remaining selected projects would be deferred to the next round.

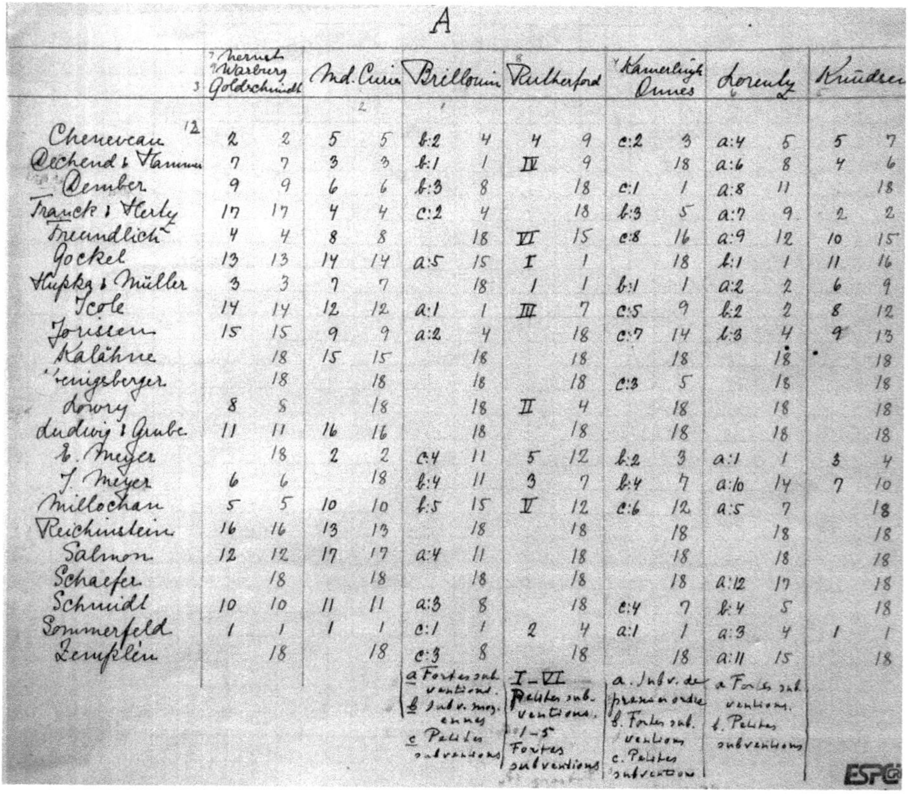

A

	Nernst Warburg Goldschmidt		Md. Curie		Brillouin		Rutherford		Kamerlingh Onnes		Lorentz		Knudsen	
Cheneveau	2	2	5	5	b:2	4	4	9	c:2	3	a:4	5	5	7
Dechend & Hammer	7	7	3	3	b:1	1	IV	9		18	a:6	8	4	6
Dember	9	9	6	6	b:3	8		18	c:1	1	a:8	11		18
Franck & Hertz	17	17	4	4	c:2	4		18	b:3	5	a:7	9	2	2
Freundlich	4	4	8	8		18	VI	15	c:8	16	a:9	12	10	15
Gockel	13	13	14	14	a:5	15	I	1		18	b:1	1	11	16
Hupka & Müller	3	3	7	7		18	1	1	b:1	1	a:2	2	6	9
Icole	14	14	12	12	a:1	1	III	7	c:5	9	b:2	2	8	12
Jorissen	15	15	9	9	a:2	4		18	c:7	14	b:3	4	9	13
Kalähne		18	15	15		18		18		18		18		18
Koenigsberger		18		18		18		18	c:3	5		18		18
Lowry	8	8		18		18	II	4		18		18		18
Ludwig & Grube	11	11	16	16		18		18		18		18		18
E. Meyer		18	2	2	c:4	11	5	12	b:2	3	a:1	1	3	4
J. Meyer	6	6		18	b:4	11	3	7	b:4	7	a:10	14	7	10
Millochau	5	5	10	10	b:5	15	V	12	c:6	12	a:5	7		18
Reichinstein	16	16	13	13		18		18		18		18		18
Salmon	12	12	17	17	a:4	11		18		18		18		18
Schaefer		18		18		18		18		18	a:12	17		18
Schmidt	10	10	11	11	a:3	8		18	c:4	7	b:4	5		18
Sommerfeld	1	1	1	1	c:1	1	2	4	a:1	1	a:3	4	1	1
Zemplén		18		18	c:3	8		18		18	a:11	15		18
					a Fortes subventions. b Subv. moyennes c Petites subventions		I–VI Petites subventions. 1–5 Fortes subventions		a. Subv. de premier ordre b. Fortes subventions c. Petites subventions		a Fortes subventions. b. Petites subventions			

FIGURE 37: Table A. Ranking by Lorentz of 22 applications examined in 1913 by the ISC (22 of the 44 requests were rejected). Numbers in black: priorities granted by ISC members; numbers in red (at the right-hand side of each column): final priorities. Thanks to a new donation from Solvay, ISIP could honour 15 requests instead of 8. Result: Franck and Hertz, in tenth position, obtained a Solvay grant for a research project which earned them a Nobel Prize for physics in 1925. Courtesy of ESPCI-PSL, Paris, Centre des ressources historiques (fonds Langevin).

The challenge was considerable. Assessing the interest of a research project wasn't easy: it required taking into account the comments of each ISC member. Evaluation had to be done in a concern of perfect impartiality, an obviously crucial requirement, but likely to complicate Lorentz's task.

Regarding the rankings established during ISC meetings, the minutes don't tell us much about the method adopted. However, we have clear information regarding the requests of the beginning of 1913, which were processed by mail. The letters attest to Lorentz's careful handling of divergent opinions, and to his taking into account of valuable advice. Here is a note, sent to him by Brillouin[433]:

"*I believe that we must first think of isolated researchers, for which a small grant can make it possible to continue work undertaken with almost no resources, even if their research deviates slightly from the one we intend to discuss more particularly in our meetings. For large grants, on the contrary, especially when they are intended for an official laboratory, I believe that we must stay strictly within the limits of our program. Thus, it seems to me better to postpone to October the subsidy requested by Messrs. Hupka and Müller...*".

And here is Rutherford's opinion on some cases at hand:

"*I would like to say that I consider the work proposed by Hupka and Müller to be of the first importance* (underlined), *not only on the theoretical side, but also on the experimental side. I think it is a matter of pressing importance at the present time to devise electrical machines to give the highest possible voltage. I have read Hupka's paper on the mass of the electron, and consider it shows evidence that he is a very able young fellow and capable of first-class work.*

Of the smaller grants, I think Gockel should be considered and also Lowry. I notice that the referee concluded that Lowry's application should be postponed. I have no objection to this; but I should like to point out in support of Mr. Lowry that he is a young man who is doing important experimental work under difficult conditions, and has to obtain his apparatus by means of outside grants.

In regard to the application of Prof. Edgar Meyer, I am in general agreement. I think the work he proposes important and very difficult, and I can quite understand the difficulty of a small Department in obtaining the expensive string electrometer and photographic apparatus required for the investigation...".

Comparison of the two reactions shows the complexity of the Committee's task (a difficulty well known to members of current scientific commissions). Confronted to the problem, Lorentz opted for an "objective" ranking procedure:

— Committee members were asked to take note of all comments on a given request, to rank the projects in order of priority, and to transmit their ranking to Lorentz. It was his task to fix the ISC order of priority, a procedure which involved certain choices. Thus, when a project hadn't been evaluated at all by a member, Lorentz gave it the lowest priority: 18. When the same order of priority was proposed by several members, the order received a weight equal to the number of the members involved (example: an order proposed by Nernst, Warburg and Goldschmidt obtained the weight 3). The priorities given by the ISC members (or by a group, such as Nernst-Warburg-Goldschmidt) were collected in a table (see the table of figure 37).

— Lorentz readjusted the individual priorities on the basis of certain criteria, such as the financial burden of the planned research. He thus obtained the ratings indicated in red. With each request, he associated a rating (a number equal to the sum of the adjusted individual priorities).

— From these results, Lorentz established the "ISC classification" of requests, starting with the one that had obtained the lowest score (*i.e.*, the highest priority).

Application of this procedure led to the selection of 15 projects, in the following order of priority:

Sommerfeld (15), Chéneveau (39), Hupka & Müller (47), Von Dechend & Hammer (66), Icole (85), Julius Meyer (87), Edgar Meyer (87), Dember (89), Millochau (89), Franck & Hertz (93), Freundlich (96), G. C. Schmidt (97), Gockel (104), Jorissen (107), Lowry (118).

Only eight of them (1-8) could be honoured within ISIP's first year's budget of 17 500 francs (the 4 000 francs allocated to Sommerfeld would allow Friedrich[434] to continue his research on X-ray interference and diffraction, phenomena that would be discussed a few months later at the second Physics Council).

This first list was submitted to the Administrative Commission on July 17, 1913. Thanks to Tassel's proposal, a second list of projects, all of which had been judged very favourably by the ISC, was submitted to the Commission.

Lorentz attached the two lists (with the beneficiaries' names and addresses, the planned research and the proposed subsidies) to a letter[435] which he sent to Tassel, with the following comments:

> "*In these circumstances, the Committee would be very happy if Mr. Solvay would consider the possibility of increasing to some extent the amount available this time. If after all he has already done for the Physics Institute, Mr. Solvay had this new burst of generosity, we would strongly recommend the physicists 9-15 on the list, for the amounts shown in the table. We kindly ask you to convey this wish to Mr. Solvay, with the message that the nature of the submitted research projects made a very favourable impression on the Committee. From the experience of this first*

year, we are convinced that the Institute is on the right track and that it will have a very beneficial influence on the progress of Science."

Lorentz's request was immediately transmitted to Solvay, who replied[436] from his vacation spot (figure 38):

"*My dear Tassel, I agree with the extra 7 800 francs*[437] *for subsidies to be allocated this year by the International Physics Institute*".

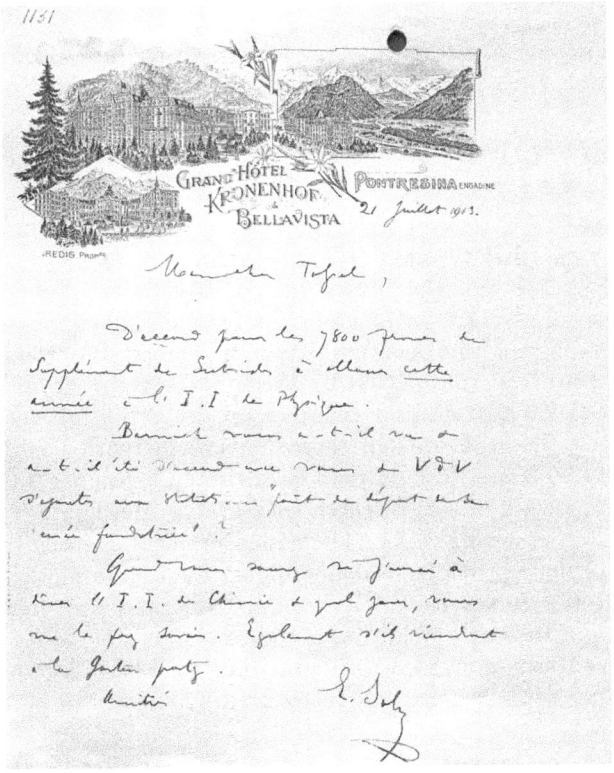

FIGURE 38: Letter from Solvay to Tassel of July 21, 1913, S.a.b.ULB. Courtesy of the International Solvay Institutes for Physics and Chemistry.

Tassel broke the news to Lorentz[438]. He asked the Committee to inform the beneficiaries that they would receive a cheque early in August.

From the forty-four requests examined during this first round, fifteen were honoured, seven were postponed and twenty-two were rejected.

The ISC members would write reports on later requests, to be examined on the occasion of future Committee meetings. Priority rankings were established on the basis of these reports.

7.3 Some notable successes

Inspection of the list of projects that benefitted from ISIP's support shows that the choices by the ISC were rather successful. In several cases, the Solvay grants contributed to a major discovery, crowned by a Nobel Prize. The following cases deserve special attention.

Max (von) Laue and Charles G. Barkla (grant in 1912)

Laue thanked the Solvay Institute in an article on X-ray diffraction, published in March 1913 in *Annalen der Physik*. He won the Nobel Prize for physics in 1914.

Barkla expressed his gratitude in an article of the *Philosophical Magazine*, published in June 1913. He received a Nobel Prize in 1917.

James Franck and Gustav Hertz (grant in 1913)

Among the seven research projects of the second list, supported thanks to Solvay's intervention, the one submitted by Franck and Hertz provides a proof of the historical role played by ISIP's subsidy program. These physicists requested 2 500 francs for research on the action of electrons on the atoms of a gas. The ISC ranked their demand in tenth position and allocated them 2 000 francs (a sum paid to them[439] on August 9, 1913). It was within the framework of this research that Franck and Hertz sent a beam of fast electrons on mercury vapor. Their measurements on the resulting electron-atom collisions provided crucial support for Bohr's model of the atom. It is therefore fair to say that Solvay's burst of generosity contributed to a major success of the early quantum theory.

Franck and Hertz expressed their gratitude to ISIP in a paper[440] of 1914. The importance of their work was recognized by the award of a Nobel Prize in 1925.

The fate of the Franck & Hertz project is a striking illustration of the unpredictability of success in physics. The eminent ISC members were unable to predict that the project deserved to be ranked in first position on their priority list. They nevertheless showed wisdom in deciding to support it.

William Lawrence Bragg (grant in 1914)

W. L. Bragg's subsidy contributed to the development, with his father William Henry Bragg, of a technique which opened unexpected avenues. Their analysis of crystal structures by means of X-ray diffraction, worth of a joint Nobel Prize in 1915, provided the basis for the discovery in 1953 of the three-dimensional structure of DNA.

Henry Moseley (grant in 1913)

Another remarkable case was that of H. Moseley, who requested £ 50 for *"research on the deflection of X-rays"*. The project having been approved by Rutherford, the ISC proposed to grant Moseley a subsidy of 1 000 francs. ISIP's Administrative Commission agreed, and brought him the news on November 17, 1913.

Historians of science emphasize the role played by Moseley's work in the acceptance of the Rutherford-Bohr atomic model. Bohr himself declared many years later[441]:

> *"The Rutherford work was (at first) not taken seriously... There was no mention of it in any place. The great change came from Moseley."*

In 1913, Moseley was going through a difficult period. The Solvay grant enabled him to continue his work[442]. ISIP's support of the Moseley project can therefore be regarded as another direct contribution to atomic quantum theory.

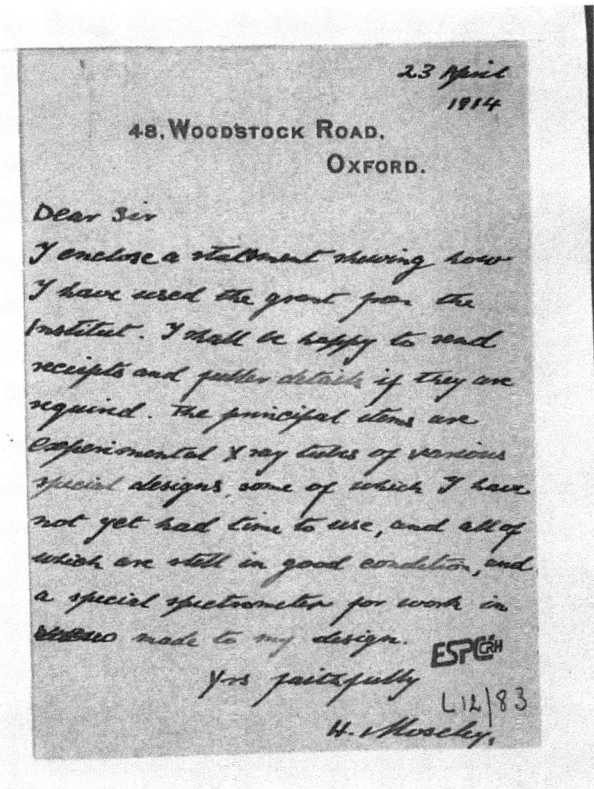

FIGURE 39: Letter from Moseley to ISIP's Administrative Commission of April 23, 1914, page 1. Courtesy of ESPCI-PSL, Paris, Centre des ressources historiques (fonds Langevin).

The case is instructive for another reason: Moseley was proposed in 1915 for a Nobel Prize in chemistry, on the basis of a report written by Arrhenius. The report made a strong impression on the members of the Nobel Chemistry Committee. Yet, they decided that Moseley was young, and that his Prize could wait. They didn't expect that this brilliant scientist would enlist in the British army, and that he would die[443] at Gallipoli on August 10, 1915.

ISIP's archives contain three documents which should be included in the file of this outstanding researcher:

- The acknowledgment[444] of receipt of a cheque of £ 39.59 sent by Moseley on December 3, 1913, to the Institute's Administrative Commission.
- A report of April 3, 1914, in which Moseley gives an account of the use he has made of the Solvay grant[445] (figures 39 and 40).
- An excerpt from Moseley's *Philosophical Magazine* article[446] of 1914, in which he thanks the ISIP for the given support (this is the article quoted by Arrhenius in his recommendation in favour of Moseley).

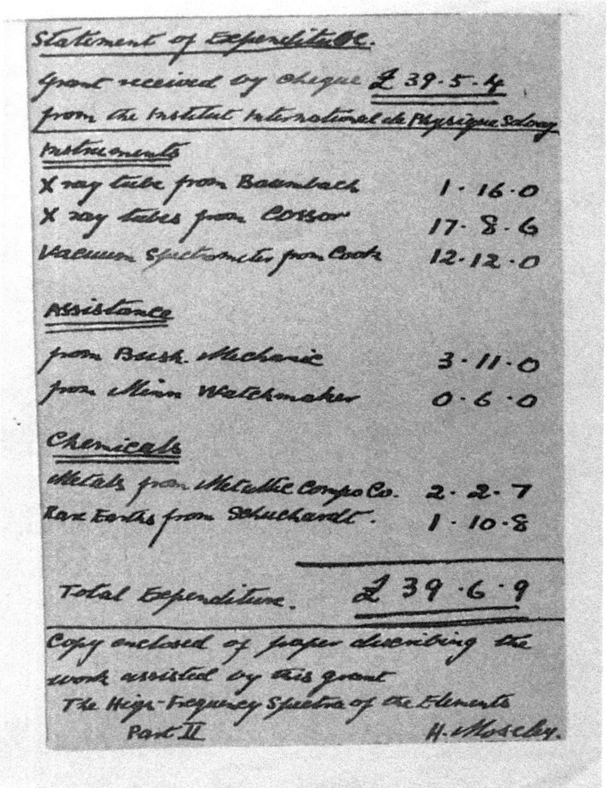

Statement of Expenditure.
Grant received by cheque £ 39.5.4
from the Institut International de Physique Solvay
Instruments
X ray tube from Baumbach 1.16.0
X ray tubes from Cossor 17.8.6
Vacuum Spectrometer from Cook 12.12.0
Assistance
from Bush. Mechanic 3.11.0
from Mann Watchmaker 0.6.0
Chemicals
Metals from Metallic Compo Co. 2.2.7
Rare Earths from Schuchardt. 1.10.8
Total Expenditure. £ 39.6.9
Copy enclosed of paper describing the work assisted by this grant
The High-Frequency Spectra of the Elements
Part II
H. Moseley.

FIGURE 40: Letter from Moseley to ISIP's Administrative Commission of April 23, 1914, page 2. Courtesy of ESPCI-PSL, Paris, Centre des ressources historiques (fonds Langevin).

Johannes Stark (grant in 1914)

Of all subsidies granted during ISIP's first years, the largest went to the discoverer of the "Stark effect". It was allocated in two steps, the last payment being made less than nine weeks before the outbreak of World War I.

Stark's attitude can be seen as an illustration of the euphoria that seizes ambitious researchers, aware of having made a major discovery. Let us briefly recall the nature of his findings and the circumstances in which he obtained ISIP's generous support.

Unlike the Zeeman effect of 1896 (decomposition of a spectral line under the influence of a magnetic field), a similar effect under the influence of an electric field had never been observed... It was precisely such an effect that Stark discovered in his Aachen laboratory in October 1913. Galvanized by his finding, he repeated his measurements in order to be sure of his observation.

He then rushed to Göttingen and Berlin to inform his colleagues[447], who confirmed the importance of the discovery.

On his return trip, Stark spent a night in Hanover and sent a four-page letter to Lorentz with the request of a subsidy[448]. In this letter, which contained tiny figures representing decompositions of spectral lines, Stark reported the positive reactions from his Berlin colleagues. He also mentioned his intention to extend his measurements and to increase their precision. For this, he needed 20 000 marks (all costs being carefully indicated): the equivalent of ISIP's yearly subsidy budget.

Stark told Lorentz that in case of success he would add a note to his Berlin report, thanking Solvay for the support given to his future research. He also planned a one-day trip, way and back from Aachen to Brussels, to visit Solvay and to show him his spectrograms.

Lorentz, who had known Stark since 1907, congratulated him for his discovery, but made him understand that it was up to the Scientific Committee to examine requests for subsidies, and that he wouldn't get anything by contacting Solvay. This warning had an unexpected effect: Stark replied that he didn't wish to obstruct proposals from other applicants, and that he wouldn't apply for a grant.

On December 16, 1913, Lorentz sent a new letter to Stark, specifying that he would plead in his favour with the Committee members. He told him that he would soon get an invitation from the students at the University of Leiden to present his discovery.

The invitation reached him through Ehrenfest. Stark's talk was scheduled[449] for February 18, 1914. Meanwhile Lorentz looked for a way to grant Stark a meaningful support. On February 5, he informed Tassel of the situation[450]:

"Stark's request caused us some embarrassment (...). I advised Mr. Stark, while leaving him entirely free, not to address Mr. Solvay, or in any case not to do so before he had received our response. He wrote to me that he had abandoned his plan to go see Mr. Solvay. It turned out that all

members of the Scientific Committee shared my opinion on the importance of Mr. Stark's research. Unfortunately, our resources were almost exhausted. We already resolved to offer 1,000 francs to Mr. Laue if he needed it (which could be the case), and in our meeting of the last Council we had such a favourable impression of Mr. Bragg junior's work that it was impossible not to recommend his request, which falls entirely within the Council's framework, and which, moreover, is rather modest. Thus, there was only 1,500 francs left for Mr. Stark. But the desires of this physicist went much further (...). It seemed to us that 6,000 francs would be the least we could offer to Mr. Stark, and that is why we have already decided to recommend him in May for a new subsidy of 4,500 francs.

By this you can see that for the next round (1914-1915) we will have no difficulty in finding a very useful plan for the money available... In October, we rejected quite a number of requests that seemed of secondary importance, but there are others that have been postponed, and on which we will have to decide next summer...

Mrs. Curie already wrote to me about the newly founded Institute of Radiology in Warsaw, which, in her opinion, deserves a subsidy. The State does nothing for it, and the Institute depends almost entirely on contributions from private individuals.

Finally, Mr. Wien from Würzburg, with whom I had some correspondence two years ago, returned to the charge. He explained to me how much a subsidy, if it could be granted without harming other interests, would be useful to him for the continuation of his research on positive rays...".

Lorentz's plea in favour of Stark proved successful. A first cheque of 1 500 francs was sent to him on February 9, 1914. Lorentz who attended his Leiden lecture, informed Stark of the Committee's proposal to grant him another 4 500 francs (taken from ISIP's budget for 1914-1915). The Administrative Commission accepted the proposal on February 24[451]. The second Solvay subsidy was paid to Stark on May 30, 1914.

The above story is revealing of another striking factor: ISIP's total independence in the allocation of Solvay grants. Stark couldn't imagine that the patron who financed all grants wouldn't intervene in the allocation process, even in the case of research of exceptional importance.

7.4 Last measures taken by ISC before the debacle

Other measures were taken during the spring of 1914 to respond to recent requests. On May 22, Tassel informed Lorentz[452] on ISIP's subsidy budget:

"The Administrative Commission recently met to fix the budget for the third financial year of our Institute. The amount to be deducted from the annuity and made available to the Scientific Committee to be distributed

in subsidies has been fixed at 19,000 francs. The Commission also decided to transfer to the subsidies a sum of 5,000 francs which hasn't been used in scholarships during the previous financial year, as well as a sum of 1,000 francs coming from bank interests. The total sum made available to the Committee for the financial year 1914-1915 will therefore amount to 25,000 francs."

Lorentz replied on May 24 that the Committee would examine the latest requests during its next meeting, which should take place[453] between June 21 and July 11, 1914. Six Committee members – Lorentz, Marie Curie, Brillouin, Warburg, Nernst, and Knudsen – met in Brussels on June 24.

The chairman gave an account of the reports written by beneficiaries of previous subsidies. The secretary was asked to send a reminder to the latecomers.

The Committee decided to publish a list of physicists who had been granted a Solvay subsidy and to inform new applicants that they could send their requests at any time to President Lorentz. Again, it was specified that support would only be granted to projects which aimed at research on molecular theory, on quanta, or on radiations.

Stark's second subsidy having been granted, Lorentz announced that 20 500 francs were available for the year 1914–1915.

IMPACT OF THE GREAT WAR

Chapter 8
The Physics Institute survives the storm

8.1 First reactions to the invasion of Belgium

On August 8, 1914, Lorentz expressed his sympathy to Solvay. A few days later, he sent the following message[454] to Héger:

> "... Also two words to tell you that I am constantly thinking of your country and its valiant defence against an attack for which it had not given the slightest cause. I hoped that we would not see a war that we expected since long, and now we see that it began in the most horrendous way one could have imagined, with the violation of the rights of a free people who were on good terms with all nations... But there is no point in talking about it...".

Lorentz followed the situation closely. Worried by the situation of Professor Wladyslaw Natanson from the University of Krakow, who was trapped on the Belgian coast, he appealed to Solvay's kindness to come to his aid. On September 12, 1914, he received the following lines from Brillouin:

> "The political procedures of the German government, which English opinion has treated with the contempt they deserve, will make it difficult for an international meeting, even a purely scientific one, to take place for years to come. For my part, it will be impossible to consider as colleagues the German scientists who did not protest against the violation of Belgium's neutrality guaranteed by Germany, and against the destruction of Louvain, as soon as they got informed, that is to say, at the moment they still believed to be victorious thanks to this treachery.
>
> No echo of such protest from German intellectual circles has reached us so far. But I do not want to judge them without knowing. Perhaps, in your neutral country, you have heard of some protests, either from

Academies or Universities as a whole, or from individual professors. It would be a consolation for me to know that some still have a sense of rights and humanity, despite German pride. In a few months, when their armies are crushed, it will be too late. It will no longer be an unselfish feeling that will make them talk.

I recognize that it takes a certain civil courage to protest against the moral failures of one's own country. In France we have the right to claim it from others. Academics and artists – not to mention other citizens – have always believed that our culture imposed a duty on us, not only towards our homeland, but towards humanity as a whole. In various circumstances, recent enough to be remembered, we did not hesitate to expose ourselves to violent retaliation for upholding the law – even in favour[455] of a single man. It is this thirst for law that is common to us, French people, whatever our political opinion. It is this thirst which gives us all the same feeling of revolt when we feel it violated elsewhere. We never dreamed of making Italy responsible for the murder of President Carnot. You can therefore see why the ultimatum of Austria to Serbia revolted us. ... And when we learned that Serbia accepted almost all conditions, we were close to despising this little people.

A people, like a man, must accept the risk of death for his independence or for an idea. This is what Belgium has done so admirably.

I now come back to the essence of my letter. Have you heard of any public or private protest, by German scholars that we know, against the violation of Belgium's neutrality, against the burning of Louvain? Whatever admiration I retain for their intelligence and for their working power, I need to know if they are of the same moral race as us: if they know how to dissociate themselves from the acts of a government without loyalty...".

A few weeks later, it was Wien – a member of the other camp – who made his point of view known[456] to Lorentz. As editor-in-chief of *Annalen der Physik*, this participant to Solvay Councils and beneficiary of Solvay grants, announced his wish to publish papers by physicists from neutral countries[457]:

"*As for the war*, he said, *we can only deplore it. I know that it will be impossible for me to forge from now on personal links with this England which I held in such high esteem. As for our good and cordial "Solvay family", it seems to me that it has forever busted apart...".*

Lorentz didn't wait for Brillouin's message to speak publicly about the conflict.

Responding to the request of the weekly magazine *De Amsterdammer*, he was eager to react to the tragedy in Louvain. In the September 13 edition, he reminded that no war could be civilized, and that it was the duty of historians to judge severely those responsible of abuses after having carried out in-depth investigations.

Then, addressing a more personal matter, Lorentz wrote: "*I feel the need these days to speak out publicly about a man whom I revere and whom I wish to honour as one of the noblest Belgian citizens.*" In a vibrant tribute to Ernest Solvay, he insisted on the man's actions in favour of science. He recalled that these actions were founded on the conviction that knowledge of the laws of nature and of societies would contribute to the happiness of the human race.

Lorentz also dwelled on the Physics Councils and on the International Institute for Physics, emphasizing its concern not to privilege any nation, and pointing out that subsidies had been granted on all sides. Even that a large part had gone to German physicists, because of their large number. "*This situation,* he said, *never aroused animosity among Belgian scientists, nor affected in any way their friendship towards Germany.*"

As President of the International Scientific Committee, prevented by circumstances from consulting with his colleagues, Lorentz felt the need "*to pay homage to Solvay, and to express his sympathy for a people so hard hit and so admirably represented by this great man*". Driven by his feelings for Belgium's citizens, Lorentz wished to honour them with this last statement: "*A people who seems to me destined to play a great role in science and civilization, and for whom I dare to hope that temporary disasters will not be an obstacle to the accomplishment of its mission.*"

Lorentz's tribute to Solvay was published in France and in Germany. The German version (from which the last sentence above was removed on Warburg's advice[458]) appeared in *Die Naturwissenschaften* on November 20, 1914.

Planck reacted to the article on November 28. He told Lorentz[459]:

"*I thought a lot about Solvay during those critical days of the capture of Brussels. The unforgettable memory of our first physicists' congress in 1911 made me very sad. According to the information I was able to obtain, Solvay's freedom of movement has now been restored, otherwise I would have insisted that the Academy, of which he is a corresponding member, should intervene in his favour, because the grain that he has sowed in his enthusiasm for science cannot be destroyed by war, nor should it be affected by our patriotism...*".

Sommerfeld, on the contrary disapproved of Lorentz's initiative[460], and made it known[461] to Wien. Reacting to Lorentz's remark in the article that Belgium had promising scientists, he declared:

"*Who is Lorentz thinking of in saying this? Would he also not care much about the truth when he sides with the Belgians against the Germans?...*".

Sommerfeld also attacked Berliner, the editor-in-chief of *Die Naturwissenschaften*:

"*To this one it is important to make clear that it does not befit a German to publish an article to the glory of Belgium.*".

Stark shared Sommerfeld's opinion[462], but didn't tell Lorentz. On the other hand, he reproached him[463] for lacking objectivity, and expressed surprise at the *"little sympathy shown towards the Germans after having maintained such a long-standing friendship with them"*.

8.2 The Manifesto of the 93

On October 4, 1914, a German pamphlet entitled *"Appeal to civilized Nations"* was spread in ten languages. It was signed by 93 German intellectuals. Among them: Nernst, Ostwald, Planck and Wien.

This was the signatories' claim:

"As representatives of German science and art, we hereby protest to the civilized world against the lies and calumnies with which our enemies are endeavouring to stain the honour of Germany in her hard struggle for existence – in a struggle that had been forced on her.

The iron mouth of events has proved the untruth of the fictitious German defeats; consequently, misrepresentation and calumny are all the more eagerly at work. As heralds of truth we raise our voices against these.

It is not true that Germany is guilty of having caused this war. Neither the people, the Government, nor the Kaiser, wanted war. Germany did her utmost to prevent it; for this assertion the world has documental proof. Often enough during the twenty-six years of his reign has Wilhelm II shown himself to be the upholder of peace, and often enough has this fact been acknowledged by our opponents. Nay, even the Kaiser, whom they now dare to call an Attila, has been ridiculed by them for years, because of his steadfast endeavours to maintain universal peace. Not till a numerical superiority, which has been lying in wait on the frontiers, assailed us did the whole nation rise to a man.

It is not true that we trespassed in neutral Belgium. It has been proved that France and England had resolved on such a trespass, and it has likewise been proved that Belgium had agreed to their doing so. It would have been suicide on our part not to have pre-empted this.

It is not true that the life and property of a single Belgian citizen was injured by our soldiers without the bitterest self-defence having made it necessary; for again and again, notwithstanding repeated threats, the citizens lay in ambush, shooting at the troops out of houses, mutilating the wounded and murdering in cold blood the medical men while they were doing their Samaritan work. There can be no baser abuse that the suppression of these crimes with the view of letting the Germans appear to be criminals, only for having justly punished these assassins for their wicked deeds.

It is not true that our troops treated Louvain brutally. Furious inhabitants having treacherously fallen upon them in their quarters, our troops with aching hearts were obliged to fire a part of the town as punishment. The greatest part of Louvain has been preserved. The famous Town Hall stands quite intact; for at a great self-sacrifice our soldiers saved it from destruction by the flames.

Every German would of course greatly regret if in the course of this terrible war any works of art should already have been destroyed or be destroyed at some future time, but inasmuch as in our great love for art we cannot be surpassed by any other nation, in the same degree we must decidedly refuse to buy a German defeat at the cost of saving a work of art.

It is not true that our warfare pays no respect to international laws. It knows no undisciplined cruelty. But in the east the earth is saturated with the blood of women and children unmercifully butchered by the wild Russian troops, and in the west dumdum bullets mutilate the breasts of our soldiers. Those who have allied themselves with Russians and Serbians, and present such a shameful scene to the world as that of inciting Mongolians and negroes against the white race, have no right whatever to call themselves upholders of civilization.

It is not true that the combat against our so-called militarism is not a combat against our civilization, as our enemies hypocritically pretend it is. Were it not for German militarism, German civilization would long since have been extirpated. For its protection it arose in a land which for centuries has been plagued by bands of robbers as no other land has been. The German Army and the German people are one and today this consciousness fraternizes 70,000,000 Germans of all ranks, positions, and parties being one.

We cannot wrest the poisonous weapon – the lie – out of the hands of our enemies. All we can do is to proclaim to all the world that our enemies are giving false witness against us. You, who know us, who with us have protected the most holy possessions of man, we call to you: Have faith in us. Believe that we shall carry on this war to the end as a civilized nation, to whom the legacy of a Goethe, a Beethoven, and a Kant, is just as sacred as its own hearths and homes."

The appeal provoked vigorous reactions in Belgium and abroad (heavy condemnation by the Belgian socialist leader Emile Vandervelde, strong reactions from the *Journal de Genève*, from representatives of French universities, from British intellectuals, and from the president of the Carnegie Institute in Pittsburgh, United States).

Reactions and counter-reactions heralded the start of a war between intellectuals, known as "*Krieg der Geister*", after the title of a German book[464], published in 1915.

This was Héger's reaction to the Manifesto of the 93 (letter[465] to Lorentz of November 2, 1914):

> *Most honourable Colleague,*
>
> *I am addressing you, in preference to any other, because I have absolute confidence in the righteousness of your character and in the safety of your judgments.*
>
> *That I am writing to you today, there is this other reason that you do not belong to one of the belligerent nations. I just saw the frontier roads of your country covered with fugitives to whom Holland gives the most generous welcome; yesterday I visited the Belgian wounded in Rotterdam and found them admirably cared for by your fellow countrymen. Hence, it seems natural to me to shelter my painful thoughts with you.*
>
> *I wish to record in front of you that I protest against the manifesto of German scholars, published on October 4th. I only learned about it late, because we are in Brussels like prisoners in our own country. We can neither receive newspapers, nor move freely, nor correspond.*
>
> *I have no doubt that you are familiar with this manifesto and that you already judged it severely. The signatories claim to protest in the name of science "against the lies and calumnies" with which the enemies of Germany water their homeland. In the name of science and their honour they countersign a document which itself is, first and foremost, slanderous and false.*
>
> *Germany, according to them, would not have violated Belgium's neutrality, and it would not be true that the life and property of a single Belgian citizen was injured by German soldiers, without them being pushed to do so by a bitter need of self-defence.*
>
> *When the time comes, when reality is known and when the facts are known in their true light, these assertions, despite having been countersigned by eminent men, will be recognized as monstrous. What I have seen and what I know authorize me to make this formal declaration to you. I took part in the war of 1870, I saw Bazeilles burn, and it was certainly a distressing sight, but it did not resemble the one offered to us by the recent German invasion of Belgium. Our government's report on what happened in Louvain, Aerschot and in the vicinity of Antwerp, only covers a small portion of the truth, because investigation could only cover a very small area. The signatories of the manifesto did not reject this report. They preferred to ignore a document that overwhelms them.*
>
> *No doubt that war involves inevitable horrors. It is by necessity bloodthirsty and cruel. But it hadn't hitherto offered us the image of the methodical cruelty and the madness of an entire people.*
>
> *When the time is right you will inquire: this is what I expect from your fairness.*

You will ask Bordet what happened in Francorchamps on August 5, ask Bruylandts, president of the Academy of Medicine, what he saw in Louvain, ask Hendrix, professor at the Veterinary School of Brussels, what he saw in Diest, ask Decraene, professor at the University of Liège what he saw in Sonval... You will learn how in peaceful villages, in Ethe-devant-Virton, in Andenne, in Louveignee, in Montigny-le-Tilleul, in Tamines, people were shot, houses looted, burnt, without any mercy with children and women.

You will conduct this investigation, you, in the name of science which is waiting to be cleared of the insult that the signatories of the manifesto have inflicted on it. You will then speak out. This is the prayer I address to you at the moment that the survivors of our little Belgian army are getting killed while bravely defending the last shred of our country so odiously invaded, so cruelly devastated.

You will make the use of these lines that you want. I know what I'm exposing myself to by writing them. I will return to Brussels tomorrow. If we don't meet again, let me thank you for the intellectual joys that your scientific discussions often gave me, and also – and above all – for your devoted collaboration to the beautiful work that we pursued together, jointly with Mr. Solvay. A work, which I am sure, you will continue imperturbably...

The *Appeal* also made a small number of Germans react. A counter-manifesto, written by the physiologist Georg Nicolai, circulated in Berlin from the middle of October 1914. It bore Einstein's signature. The purpose of this counter-manifesto was to alert "good Europeans", according to Goethe's expression. Yet, it provoked very few reactions.

A British response to the *Appeal* of the 93 intellectuals appeared in the *Times* of October 21, under the title *Reply to German Professors*. It was signed by 117 scholars. Among them: Lord Rayleigh, W. H. Bragg, Sir Joseph J. Thomson and Sir William Ramsay. The signatories found it hard to believe that the *Appeal* represented the opinion of German professors. They didn't doubt the sincerity of their repugnance for war and their zeal for culture, but they condemned the deliberate destruction of cultural treasures by the German army, such as Louvain and the cathedrals of Malines and Reims. On the other hand, they blamed the responsibility for the war to Germany, for it was she who violated Belgium's neutrality. Great Britain aspired to peace, but this desire didn't interfere with their duty as intellectuals, which was to establish the facts and to report on them.

In Paris, the Academy reacted on November 3, recalling that the Latin and Anglo-Saxon civilizations had made for three centuries the greatest contributions to the development of mathematics and physical sciences. It was from these cultures *"that the most important discoveries had emerged"*. The Academy protested against the claim that the intellectual future of Europe would be determined by German science. It also refuted the assertion that the salvation of European civilization depended on a victory of German militarism, in solidarity with German culture[466].

Among neutral nations, an initiative for a "*Europe at peace*" was launched in 1915 by Dutch pacifists. Their idea was – once peace negotiations would begin – to address a petition to the leaders and elected officials of European states. This petition, signed by ten scientists per country, would be a plea for a reconciliation between citizens of belligerent countries. Its text suggested to forget all that separated and divided. It recommended forgiveness for injustices and abuses committed by the nations at war.

The request for signatures should be sent to specific scientists, once the final text had been agreed upon by the pacifists.

Lorentz accepted to sign the document after some modifications. He and Zeeman provided a list of scientists likely to follow their example. The signatories exhorted the leaders "*to show themselves worthy of their mission, i.e., to safeguard the destiny of humanity by using their influence in a spirit of sincere reconciliation, in order to bring about a peace that wouldn't contain any germ of a new conflict, and to unite the divided European nations, so as to guarantee the freedom and prosperity of all in a concern of perfect equality*".

The petition received only a small number of signatures, but it raised a wave of protests, especially from Voigt[467] and from Brillouin. This is how the latter reacted in a letter[468] to Lorentz of February 25, 1915:

Dear Sir,

I hasten to reply in the negative (underlined) *to your circular, which, I confess, at first made me very angry, and then caused me some pain.*

I believed that the fate of our unhappy countries deserved more sympathy. We do not want to forgive or forget, we have nothing to apologize for.

Did you realize on this date of February 1915, that the signature of a citizen of the Allied peoples at the bottom of your petition would be equivalent to the gesture of kneeling before the Kayser, raising hands to heaven and asking for mercy.

It is a gesture that does not suit me, and I know of no way to make me do it as long as my wife and my daughter are safe. I doubt that you will get a single signature in France, in Belgium and in England, except perhaps that of the discredited figure named Caillaux[469].

Belgium is still invaded, oppressed and treated as a conquered country. Reims is bombed every day. A fifth of France has been systematically burned, looted, devastated... The most peaceful inhabitants taken as shields by the German troops. This almost ceased to amaze us after so many testimonies, but we now know that there will be no security in front of the German nation unless we put a strong muzzle on it. We will work, as long and as energetically as it takes, to build this instrument of international security, or we will perish.

Your petition comes several months too late, or too early. I, at least, hoped until the first Moroccan affair, that the Germanic nation was not an accomplice to its unbridled militarism, that there would perhaps come a day when the nation would free itself from its masters, and when without ceasing to have disturbing passions for us, this nation would begin to understand the word "fraternity" other than a certain Cain.

I lost this illusion by chatting[470] *with Mr. Wien.*

As for not including in the peace treaty, that we will hopefully dictate, clauses that would be the seeds of war, I do wish that it will be the case... I learned what German intelligence is worth in the sciences. I don't accept any assimilation of our compatriots with the Germans. You have known for several months what haughty contempt I have for the signatories of the Appeal to the civilized nations...

8.3 A dead end conflict

The Great War caused an exodus in universities and research centers. In April 1915, the *Physikalische Zeitschrift* listed 150 German or Austrian physicists called up for the service. Among them, 119 were postdocs and 31 were professors. Obituaries followed one another. People soon learned that Fritz Hasenöhrl, one of the Austrian members of the 1911 Council, had been killed on the Italian front in October 1915.

Back from a trip to New Zealand, Rutherford sent this message to Lorentz (letter of June 7, 1915):

My dear Professor Lorentz,

I have not heard from you except indirectly for nearly a year. You may have heard that my wife and I and our daughter went to Australia in July to attend the Meeting of the British Association, and saw my people in New Zealand.

We came back via Vancouver, Montreal, New York, and arrived in Liverpool on January 7th. Fortunately we had no special adventures by the way, and we got quite accustomed to travel with lights out by night. I saw many of my old friends in Montreal. Since my return I have been busy with my University work, and continuing my researches.

My laboratory is, of course, like all others, practically depleted of research students, and the number of senior students is much reduced, owing to enlistment.

A number of my lecturers and researchers have Commissions. Of these, Darwin and Andrade are now in France, while Florance, Pring, Robinson, Moseley and Walmsley are in training for the front. I expect you have heard that the Universities have provided a considerable number of officers for the new Armies. Manchester has contributed over 400.

I went down to London last week to give a lecture at the Royal Institution on "Radiations from Exploding Atoms", and had a good audience. I saw Lindemann there. He tells me that he escaped just in time from Germany, and is now at work in the Air-craft factory near London.

We have a number of Belgian refugees in our midst, including several University Professors and others. They tell me that Verschaffelt is now teaching in Leiden. I am very glad to hear that he got away safely and has got employment.

I hear very conflicting rumours about Solvay, who, I understand, is still in Brussels. I am afraid that whatever the result of this war, it will be difficult to have our Solvay meetings again for some time. It is very unfortunate, but it looks to me that the methods of German warfare are such that the allies will have no social intercourse with them for a long time to come. All the evidence points in that direction. Germany now seems to regard the promiscuous murder of non-combatants as one of her chief methods of warfare, and everybody is expecting her to adopt still more barbarous methods than the use of poisonous gases[471], poisoning of water etc. in the future. The German populace appear to be educated to believe that all methods are good in war that led to a temporary success.

I should imagine that you have been passing through rather anxious times yourselves, and I expect you to have done all you can do to look after the great number of Belgian refugees etc. I daresay that you saw that Arrhenius and other Swedish representatives have signed a memorial protesting against the Germanic methods of warfare. I think it is one of the few things that neutral countries can do at the present time.

We have a number of Australian and New Zealand wounded who have come from the Dardanelles, and it is very interesting to hear their experiences in the terrible fighting that is going on there. It is quite common in this country to meet wounded soldiers from nearly all parts of the earth. I travelled yesterday with an officer who has taken part in the fighting in the Persian Gulf.

We have had very fine weather here, and the country is looking beautiful. Apart from the great numbers of people in khaki, it is sometimes difficult to believe that we are at war. The Universities go on much as usual, but of course with diminished numbers. Work is everywhere very plentiful, and it is only certain small classes of people who really feel the pinch of the war.

I saw Sir J. J. Thomson in Cambridge a short time ago, and found him quite well. His only son is now working at the next bench to Lindemann in the Air-craft factory. He spent a fortnight in the trenches in Flanders in the wintertime, but was sent back owing to illness.

Young Bragg is an officer in the new Army. I may tell you privately that it is quite likely that Professor Bragg will be appointed to the Chair of

Physics at University College, London, in place of Professor Trouton, who has resigned after a bad illness

I shall be glad to hear from you. When you visit Leiden, please convey my kind regards to any of my friends there, including Dr. Fokker, who, I believe, is now teaching there...

A new episode in the *Krieg der Geister* opened in 1916, following the publication in Paris of a book by historian Joseph Bédier, based on notes taken by prisoners of war and entitled "*German Crimes according to German Testimonies*".

Deeply struck by Bédier's conclusions, Lorentz shared them with Planck and Wien, asking them to find a way that would allow German scholars to distance themselves from the "*Appeal to the civilized nations*".

Such a gesture, said Lorentz, "*would be likely to facilitate the reestablishment of scientific relations after the war*".

Wien didn't want to hear anything[472]. The request seemed insane to him... Planck reacted favourably. After consulting Lorentz, he decided to send him an "open letter". These were his words:

"*I observed on several occasions, and to my great regret, that the Appeal, by reason of its wording, gave an incorrect impression of the views of those who signed it.*

My opinion is shared by others, in particular by Nernst. The Appeal, which, in the terms used, reflects the first weeks of the war, can only be understood as an act of defence: a defence of the army against the serious charges brought against it, and the claim that scientists and artists did not intend to dissociate themselves from the cause of the army.

We cannot, of course, feel responsible for every act committed by a German, and it is too early for us to have an authoritative judgment on these facts. As long as the war continues, we will have a duty to serve our country.

I wish, however, to stress that I am firmly convinced that there are moral and spiritual spheres which lie well beyond the conflicts between nations, and that an honest cooperation to preserve these universal cultural values, as well as the respect for the rights of the citizens of an enemy country, can be reconciled with the deep affection we have for our homeland, and with the energy we devote to it...".

Lorentz published Planck's letter (accompanied by a few introductory words) in the *Algemeen Handelsblad* of April 11, 1916.

The letter also appeared in the international press. Reactions were very diverse. Brillouin declared[473]: "*Planck still doesn't understand*".

But for Lorentz, the letter was a first step which pushed him to intervene with his German colleagues so that they would come to the aid of certain

Belgian prisoners, in particular Hostelet, condemned to five years of forced labour for having helped British soldiers to escape from a small village in Hainaut.

Touched by the distress of the physicist's wife, Lorentz turned[474] to Planck and to Warburg. Both intervened and obtained a relaxation of the conditions of detention of Hostelet, allowing him to continue his studies[475].

A few months later (January 1917), Lorentz joined the appeal of a dozen Dutch intellectuals to the Reich's Chancellor to demand an end to the deportation to Germany of Belgian civilians. The appeal seems to have contributed to a suspension of the systematic deportation of unemployed Belgians[476].

Contrary to Planck's and Warburg's positive reactions, the attitude of Sommerfeld raises questions[477]. This regular member of the Physics Councils, who benefitted from a Solvay subsidy, distinguished himself by actions which made him undesirable in Belgium after the war.

Let us recall the facts (reported by Sommerfeld's biographer, Michael Eckert).

1. In 1916, the German occupier intended to transform the University of Ghent into a Dutch-speaking institution. Sommerfeld consulted Kamerlingh Onnes to find out whether one of his collaborators, Willem H. Keesom, would accept to become a professor at the transformed university. Having discussed the matter with Lorentz, Onnes replied that it was up to the Belgians to appoint their professors. Sommerfeld noticed that his views weren't shared by his Dutch colleagues, but their opinion wouldn't prevent him from continuing his action.

2. In January 1918, Sommerfeld went to Tournai and lectured in physics to German soldiers. Shortly afterwards, he delivered part of his lectures at the University of Ghent, giving them a clear political dimension. His lectures were applauded. The celebrated Munich professor was welcomed with open arms.

3. Enchanted by the success of his lectures, Sommerfeld displayed his optimism in German newspapers[478] confirming his wish to support the enforcing of the use of Dutch at the Ghent's University... But that wasn't all. These are his words in an article of April 1918: "*I never believed that a forced annexation of Belgium by Germany was possible, nor that it was desirable. But the relations which now seem to be on the verge of forming in Flanders, more precisely in the cradle of Flemish culture, are likely to favour an annexation freely consented to, by the part of Belgium which is most valuable to us. It would be a shame if we were to disregard the legitimate aspirations of the best elements of this people. I left Ghent with the conviction that relations between Germany and Flanders will soon see the promise of a new future...*".

4. When the paper appeared, Sommerfeld was lecturing in Brussels[479], providing the German army with details on matters of a military nature, especially in ballistics[480].

There are indications that these lectures took place in the reading room of Solvay's Sociology Institute. We know that the Institute had been requested by the occupier as a conference room for the military (on January 31, 1918, Lorentz had asked Planck[481] to intervene in order to prevent the occupation of the Institute, but Planck's request had been ignored) We also know that Sommerfeld lectured for a small audience[482]: an element which agrees with the fact that space in the Institute's reading room was very limited.

One point seems firmly established: Sommerfeld's actions in occupied Belgium were deemed sufficiently serious to justify his exclusion from the 1927 Council, while his work on spectra amply justified his presence alongside Bohr (see section 9.3).

Nernst's secret negotiations

We mentioned Planck's open letter regarding the *Appeal of 93 German intellectuals*.

What about Nernst, other signatory of the Manifesto?

We know that the famous chemist participated in the manufacture of combat gases[483], a program developed by his colleague Fritz Haber, who tested its effectiveness in Ypres on April 22, 1915.

What is less known is that *Geheimrat* Nernst was put in charge of a secret mission. Its purpose was to enter into contact with Goldschmidt's father-in-law, the influential banker Franz Philippson. The mission and Philippson's attitude have been described in detail by the banker's biographers R. Brion and J.-L. Moreau[484]. We derived the following elements from their book:

— May 1915: Nernst went to Brussels with a mandate from the German government. His purpose was to see Philippson. The idea was to get in touch with King Albert in order to obtain his adherence to a separate peace project. Details of the project haven't been revealed, but it is clear that Philippson didn't accept Nernst's initial request.

— June 1916: Far from giving up, Nernst returned to Brussels with a new proposal. His goal was to involve Philippson in a general peace project that would be negotiated on neutral ground between belligerents on the western front. The idea was to obtain the banker's personal intervention. Philippson would transmit the German Chancellor's proposal to the Allied governments through the King of the Belgians. Nernst would get him passports to go to Le Havre. Belgium would obtain the guarantee of a return to its pre-conflict position... The banker hesitated: should he warn the government? Was he not in danger of being accused of collaborating with the enemy? His first reaction was to write a letter to the Belgian government in exile, asking if he should go to Le Havre. He finally decided not to send the letter[485], the aftermath of the Somme offensive having ruled out any possibility of opening talks... In search of support, Philippson contacted the Belgian statesman Emile Francqui and informed him of Nernst's approach. Francqui transmitted the information to Herbert Hoover[486], his American counterpart in London, who headed the *Commission for Relief in Belgium*. The latter got in touch with the British government. Informed in turn, the French government announced that it would oppose any peace proposal. Result: Philippson was asked not to deal with the matter any longer, and was officially denied any access to the King.

— November 1916: Nernst visited the banker for a third time. His message was clear: the Belgian population was starving and threatened with deportation; the country's industrial apparatus was in the process of being dismantled... If the government wanted to avoid the total ruin of Belgium, it had to encourage the Allies to enter negotiations as quickly as possible. Wasted effort: Philippson had learned his lesson: he refused to intervene.

— December 1917: Nernst went to Brussels for a last time. He told Philippson:

"Why does Belgium persist in prolonging a war which doesn't concern her? Can she not put pressure on Great Britain so that it stands out from the hard-line attitude of France?"

Foreseeing a favourable outcome of the war, Philippson replied that the Belgians no longer had confidence in German promises of a restoration of the independence of their country. In order to avoid any further discussion, he added:

"When you told me in June 1916 that the German government had decided to grant this independence, I was quite ready to continue our conversations. I gave up the idea the day after your departure by reading your Chancellor's speech in which he made explicit reservations on the subject... England will fight for the freedom of the world. As she did two centuries ago against Louis XIV, and one century ago against Napoleon, she will fight against Wilhelm II, with the help of the United States...".
Nernst took the banker's reaction for granted and didn't insist any longer.

One question remains open. Why did Goldschmidt's father-in-law agree to receive a visit from a man who had signed of the odious *Appeal to civilized Nations*?

We know that Goldschmidt, German born, became an ardent Belgian patriot[487], and that he remained in Paris during the war as director of the Belgian army's *inventions service*. Yet, it is not excluded that Goldschmidt pleaded "in private" the cause of his former mentor, emphasizing his good feelings towards the Belgians. He may have told his father-in-law that Nernst signed the *Manifesto* without reading it, reassured by the fact that it carried Planck's signature.

Whatever Goldschmidt's role in this affair, it is clear that Nernst was deeply disappointed by the failure of his mission. His bitterness appears in this message[488] to Lorentz of February 29, 1919:

> "In Germany, intellectuals and all those who ruled the country during the war can invoke many steps, private or public, taken to put an end to the conflict. But the Entente wouldn't hear anything, thus prolonging the war for at least two years. She rejected, as is well known, the mediation of the Pope which we had accepted. The Entente achieved its goal, the future will teach us at what cost...".

8.4 Edition of the second Solvay volume: a thankless task

Let us briefly go back to the summer of 1915, when Lorentz and the members of ISIP's Administrative Commission, learned that they could no longer correspond by letter. Henceforth only postcards would be authorized.

On June 25, 1915, Tassel sent this note[489] to the director of the Brussels printing house Hayez:

> "Professor Lorentz just returned to me the proofs for pages 1 to 64 of the Physics Council. Due to the large number of corrections he had to make to the report of Mr. J. J. Thomson, it isn't yet possible to deliver the "Permission to print" for this part of the work. I would therefore be grateful if you would make sure that the corrections indicated by Mr. Lorentz have been made, and if you could send him (Professor H. A. Lorentz, 76 Zijlweg, Haarlem, Holland) a new proof of the attached pages...".

Two days later, Lorentz sent a letter to Tassel, specifying the title page of the second Council's report:

> "Problems regarding the structure of matter. Reports and discussions of the second Physics Council, convened in Brussels from October 27 to 31 by the International Solvay Institute for Physics. Published by the Secretaries[490]".

On August 2, 1915: Bolt out of the blue: Tassel got this disturbing message[491] from Lorentz:

> "I received the new proofs (pages 1-64) of the report and wanted to send them to you, but I got them sent back to me with the words: "Zurück unzulässige Sprache (Back inadmissible language)". As there are still a number of errors, and the same will be true of the sheets which will follow, I must suggest that you wait with the printing until the proofs can be freely exchanged. I will address Mr Warburg in Berlin, who will no doubt wish that this scientific work be printed...".

Four days later, Lorentz was informed by the printer[492] of the latest developments:

"Following Mr. Tassel's advice, I turned to the German Censors who suggested to me the idea of dispatching the shipment with the help of the German Consul closest to your city... As Mr. Tassel and I wish to finish this work which had already cost so many immobilizations and costly corrections, I take the liberty, Mr. Professor, to ask you to return to me the highest quantity of proofs and manuscripts possible, of which Mr. Tassel will supervise the printing so as to reduce the number of shipments.

Above all, I urge you not to allow work to be suspended again, as a new stoppage would be highly detrimental. Alternatively, and in the event that the intervention of the German consul is not possible, I suggest that you write to Aachen's Censorship, letting them know the composition of the Committee...".

According to Hayez's request, Lorentz addressed the Imperial German Consulate General for the Netherlands, indicating the composition of the ISC and insisting on the purely scientific nature of the shipment. He explained that it had been agreed that the Council's reports and discussions would be printed in the original language (*i.e.*, partially in German, French and English) and requested the consulate administrator to intervene as an intermediary. The latter replied that he would send the documents to their destination. He attached to his reply to a copy of the letter he had sent to the Political Staff of the Governor General for Belgium... Alas, without the slightest result: the shipment didn't arrive.

On October 12, 1915, Tassel had no choice but to inform Lorentz that the proofs to be sent by the German consulate in Amsterdam, hadn't reached Brussels. On December 22, he sent this[493] message to the printer:

"As a result of our conversation yesterday, I would be grateful if you would, when acknowledging receipt of the proofs of the Council's report corrected by Mr. Lorentz, tell him on my behalf that I will check with great care the final proofs that you will send me, so as to verify that all indicated corrections have been taken care of.

Please also tell Mr. Lorentz that after finishing the printing you will keep the sheets in store and postpone the work of binding for and completing volumes until later, the moment being not particularly favourable for their distribution. Of course, a finished copy should be sent to Mr. Lorentz as soon as the work is completed...".

Copies of the volume were available from the beginning of 1916. However, Tassel wasn't at the end of his sorrows: the sending of material from Belgium to Holland turned out to be just as complicated... On March 5, he got a new distressing message from the director of the printing house Hayez[494]:

"I tried unsuccessfully to send the volume of the second Physics Council to Professor Lorentz. At the Holland legation I was refused shipment

without a proper Censorship authorization; at Censorship I was refused this authorization, and told that sending should be done by post. But the post has, I think, given sufficient proof of her irregularity not to attempt yet again to entrust her with a shipment. If, happier than me, you had a possible way of sending, I would be grateful if you would let me know. I take this opportunity, Sir, to thank you again for the complacency you have been kind enough to show to help the completion of this work, made very laborious by the circumstances...".

Tassel replied:

"Unfortunately, I do not have any means of providing Mr. Lorentz with the good sheets of the volume that you haven't been able to send him so far. I think the best would be to risk sending him two copies simply by post, as you were told at the Censorship, but by express-recommended mail. If the shipment gets lost, it doesn't matter anymore since the job has been completed...".

We don't know if the copies reached their destination. But the exercise proved to be vain for another reason. Fearing a seizure of ISIP by the German occupation, the Administrative Commission refused to submit the report to the Censorship, and requested the printer not to put the work into circulation. House Hayez was asked to keep the volumes in its store.

Soon after the Armistice, the Administrative Commission announced that the Institute's publications would be in French, in compliance with a measure adopted in 1919 on the occasion of a revision of ISIP's statutes[495].

Result: The copies of the "Hayez edition", produced at the cost of so much effort, were destroyed[496]. Yet, one of them has been preserved: it can be consulted at the Archives & Libraries Service of the Brussels' Free University... As to the official report of the second Solvay Council, edited in French, it was published in 1921 by Gauthier-Villars.

8.5 Solvay's actions and projects

Solvay participated from the beginning of the war in efforts to relieve the Belgian population trapped between the allied economic blockade and the German refusal to contribute to the supply of the occupied territories. In October 1914, he created with Dannie Heineman[497], an American engineer and businessman settled in Belgium, and with the help of Francqui, the *National Relief and Food Committee*.

And yet, the industrialist didn't renounce his scientific activities. During the first months of 1915, he started setting up an "astronomical system". His idea was to replace gravitational attraction by a *Gravito-Materialitic* theory of *"transmutation"*.

On May 20, Solvay went to Holland, not without having given instructions to Herzen for the establishment of an *"absolute Gravito-Kinetic geometry"* that should lead to a *"theory of the absolute active Universe"*.

Herzen set to work, relying on the industrialist's ideas and on the directives that were sent to him during Solvay's summer holidays in Switzerland.

Back in Brussels, Solvay took notice of the work accomplished. Disappointed by Herzen's results, he declared[498]: *"The whole Gravito-Materialitic theory is to be taken up again."* Then, taking stock of the situation in December 1915, he wrote in his diary:

"All my publications, all my institutions which followed one another, prove beyond what is necessary, that I tried to be the realizer of my scientific concepts as I had been of my industrial ideas, either by myself, or by those whom I put in a position to continue, and possibly prolong, my action, and who were willing to accept this situation... There is an interest that the work be presented after me as it would have been during my lifetime. It suffices to remember the exhortations Mr. Henry Le Chatelier[499] *made to me in this regard...".*

Solvay was clearly worried. He feared that by disappearing too soon, his collaborators wouldn't succeed in preserving unity in his life's work... However, he quickly regained his confidence, underlining the novelty of his approach, and reassuring himself with the idea that it would soon *"overthrow the old dynamic theory"*. Then, in a note of December 30, he reaffirmed his adherence to the Berthelot-Thomsen principle of "maximum work", and exposed his own views on the "degradation of energy".

Need of a new human society

Despite his unwavering confidence in the value of his *Gravito-Materialitic* approach, Solvay couldn't ignore the depth of the changes that the Great War was due to bring about. Confronted with the European catastrophe, he realized that the biggest challenge was a social one. In May 1915, he concluded that it was time to build a new science-based society. His idea was to launch a "movement of opinion", designed and supported by scientists. *"They alone, he said, have the power to restore international harmony, so severely shaken, and of which humanity has so much need."*

To achieve this goal, Solvay relied on his partners of the academic world. Sojourning in Holland, from May 20 to 30, he laid out his plan to Lorentz. Then, during his summer holidays in Switzerland, he gave funds to Ph.-A. Guye, asking him to start the "movement of opinion" in Geneva[500] as soon as the hostilities were over.

In May 1916, Solvay was again in Leiden, visiting Kamerlingh Onnes' laboratory with Tassel and Warnant, and expressing keen interest in superconductors[501]...

But his main concern, the need of a social reform conducted by physicists and chemists, hadn't changed.

Solvay's plan entered a new phase during the summer of 1916. On July 29, he responded to a letter from Emile Vandervelde, a well-known socialist and member of the Belgian government in exile[502], who asked for financial means to come to the aid of Belgian workers abroad. Solvay promised a credit of 15 000 francs and announced his plan for the construction of a new society:

> "This society will have its own science, just as mechanics and physics (...). You will see, my dear Emile, it is our glorious little homeland which is destined to introduce this new era (...). A great productive and democratic movement will necessarily succeed the current cataclysm. We will be ready, I hope, in our little Belgium, to correctly answer the future social issues...".

News from the IACS

An unexpected event happened in the beginning of August 1916. Solvay received a letter from Haller, announcing the death of Sir William Ramsay on July 23, and expressing concern about his replacement. The president of the IACS reaffirmed his position on the thorny question of the location of the Association's annual meetings:

> "As for our next meeting, I want it to take place in Brussels. I'll do my utmost to uphold this opinion, for we want to implicitly render a discreet homage to the nation which has so courageously, and in such a chivalrous way, risen to the challenge of the barbarians. In this commitment there is, moreover, an after-thought, which is that the Austro-German representatives won't have the audacity and the effrontery to go to a country that their soldiers have so despicably treated and devastated...".

Haller took advantage of his letter to announce his intention to propose, together with Ph.-A. Guye, Solvay's election as a corresponding member of the Paris Academy.

Touched by this mark of esteem, the industrialist hastened to thank Haller[503]. He also informed him of his wish to reform human society by relying on his International Institutes, an objective already communicated to Lorentz and Guye:

> "You will examine this together, because it relates to concerns at the end of your letter, quite legitimate (...). As to the choice of Brussels for the next meeting, I absolutely agree to all points of view, and precisely above all to the one I am also considering (...).
>
> My dear and good Haller, there comes a day to create an "International consultation between the representatives of the physical sciences who consider the possibility of a positive, gradual improvement of social organization". The positive law that they will draft will lead to the merciless

condemnation of the deliberate, systematical and long-term perpetrators of crimes. The law will lead to the obligation of public retraction for those who, subjected to sudden hypnosis, will have inspired to crime, and to the acquittal of those who, henceforth revolutionaries, will have acted on orders after having been previously blinded, without knowing...".

Solvay, as we can see, was willing to adopt a conciliatory attitude to those who, having signed the *Manifesto of the 93* in a moment of blindness, would have amended in time[504], an attitude similar to that of Lorentz and quite removed from the radical positions of Brillouin and Haller.

The Social Action Committee

Solvay's next care was to induce Lorentz to get in touch with Haller and Guye for the launching of his project. On August 14, 1916, he sent him the following message from Switzerland[505]:

My dear and good Lorentz,

You will find, included and confidential, some letters and their answers, which, if you can read them in order of date, will update you on the social action that I am seeking to initiate, and of which the International Institutes that I founded should be, if possible, the starting point, in accordance with the preliminaries I already mentioned to you in May 1915... Haller, Guye, and my son Armand, have copies of the same included letters, because I consider that they are exclusively part of a first Committee which should be able, in my opinion, to lead this action (...). In short, as to myself, I think – and I officially want to make it known to relieve my conscience – that the extraordinary moment which presents itself is absolutely and imperatively designated for physico-chemists, under penalty of failing in their most serious humanitarian duty, to take position in the current social mess, and this as true representatives of positive science, by establishing that physical energetics regulates the conditions of individual and social life, as well as all other conditions, and that it must henceforth be adopted as a principle of development of their progress (...).

Such is, my dear and good Lorentz, the thought that I can't help but have, in my special situation, of the action that I wish to see taken. But this without having the slightest intention of influencing the thought and the action of the three of you, other than by what I believe to be of rigorous scientific essence[506] *(...). Anyway, I open as of now, for these possible purposes, a loan of 25,000 francs in Geneva, one of the same amounts in Paris, and a similar one in Rotterdam, from which each of you can possibly draw...*

Unfortunately for Solvay, things didn't develop as expected. Driven by his desire to share his concerns with his scientific partners, the valiant industrialist underestimated the communication problems between the members of his Committee... He also seemed unaware of the difficulty for a scientist to engage in an action which drags him away from his sphere of competence.

Lorentz, as we shall see, felt unable to conceive any link between Solvay's social project and the means of action of a Physics Institute. In addition, he had been deprived by the war of all means to consult with his foreign colleagues.

Haller was equally puzzled. This is a note[507] which he entered in his diary on August 24, 1916:

> "Received a letter from Talvard, announcing that Mr. Solvay has placed at my disposal a sum of 25,000 francs to defray the expenses that might be caused by the assistance which I would give to H. A. Lorentz from Holland and to Ph. A. Guye, for the realization of the scientific-sociological ideas of the valiant industrialist.
>
> Wrote immediately to Ph. A. Guye to ask him what action Solvay asks us to perform, and in what form this action should be shown. I know exactly what could, and what should be done by the International Association of Chemical Societies, but I don't quite understand how the physical sciences can contribute to solving such complex questions which pertain to sociology. Guye and Lorentz received some ideas on the subject verbally from Mr. Solvay, and it is up to them to formulate them and to make a program. As for the money put at our disposal, I don't know what to do with it and will not touch it for the moment".

Back from his summer holidays, Solvay sought to extend his action at home. As a member of the board of the Université Libre de Bruxelles, he submitted a note *"The University of Tomorrow with Positive Physical Objective Trend"*, in which he pleaded for the creation of a special course to enlighten first-year students on the social significance of their university education. Solvay's goal was twofold: to communicate to the University the *"renovating impulses generated by the overwhelming European catastrophe"* and to link the planned *"social renewal"* to his scientific theories. Indeed, the renewal he wanted to promote had to appear as the result of his scientific investigations[508].

Dissolution of the IACS

What about Haller's statement *"I know exactly what could, and what should be done by the International Association of Chemical Societies"*?

There is little doubt that it reflected the president's desire to put an end to the Association he had created with Ostwald in April 1911. Indeed, from 1916 on, Haller persisted in his wish[509] to dissolve the IACS, largely dominated by the Germans. The next step would then be the creation of a new international organism from which all representatives of the Central Powers would be excluded.

Haller's wish was fulfilled. A proposal for dissolution of the IACS was introduced on March 14, 1919, by the French office[510], in agreement with the International Association of Academies. The measure resulted in the application of the clause provided for by Tassel: *"Restitution to Mr. Ernest Solvay of all the sums and foundations*[511] *made available to the IACS".*

Thus, finding itself freed from any commitment to Chemical Societies, ISIC's Administrative Commission succeeded in achieving Solvay's initial objective: the creation of a Chemistry Institute, modelled on ISIP.

An autonomous *International Committee*, responsible for ISIC's scientific direction, was set up in 1920. It was chaired by one of the rapporteurs to the second Physics Council: Sir William J. Pope, professor at Cambridge University, and vice-president of the International Union of Pure and Applied Chemistry (IUPAC[512]).

ISIC's Committee members met on June 24, 1921, one day before IUPAC's second conference in Brussels (the first was in Rome). Attendees were invited by Solvay on June 29 at the château de la Hulpe.

But that would be all: IUPAC wouldn't replace the IACS in the direction of Solvay's Chemistry Institute.

The first Solvay Council on Chemistry was held in April 1922, one month before Solvay's death. It was a reminder of the founding event that had taken place in 1913.

Brussels was once again the host of a historic meeting[513].

8.6 Satisfactions, setbacks and hopes

The numerous notes written by Solvay during and after the war – his diary contains more than a hundred of them – are proof that he continued his research with the firm hope of submitting his *"Fundamental synthesis of the Universe"* to the judgement of professional scientists. Impatient to finish his work, he flooded his collaborators with increasingly pressing requests.

On top of his activities at home, Solvay managed to escape on several occasions from occupied Belgium to visit neutral countries. He went to Holland with his closest collaborators, notably in May 1916[514], and spent his summer holidays in Switzerland, until the year 1918 (as we shall see).

The summer of 1917 proved beneficial. Haller kept his word: on June 18th, Solvay was elected correspondent member of the Paris Academy of Sciences, a distinction which went straight to his heart (he was a corresponding member of the Royal Prussian Academy since May 23, 1913).

The next summer was much more disappointing. Solvay did not obtain his passport for Switzerland: he was deprived of his usual vacation in the mountains.

The Physics Institute survives the storm

On August 28, 1918, his secretary Ch. Lefébure informed him[515] of the reasons of the refusal (we reproduce Solvay's annotations of the letter in parentheses):

Dear Mr. Solvay,

Yesterday I saw Mr. Deutcher, the advisor of the Swiss Legation... He asked me to tell you his sincere regrets, and to remind you of his great hope for success.
The reason for the refusal of the passport comes from Berlin itself: it was caused not by your refusal to make your Institute available to the occupier – which they could understand – but because you would have said at a time, to a person unknown to Mr. Deutcher, that even after the war any collaboration with German scientists would henceforth be impossible.

(Solvay: I never said this. Science is international before anything else, and, of course, it will universalize after the war, asserting itself more and more as representing the exact. Science will even go so far as to judge the scholars who deviate from this guideline in the notorious acts of their everyday life. This is my opinion).

The Berlin authority went to find the academicians, told them that you had made this point; these gentlemen believed it, since the information was official, and were therefore prevented from acting for you. Mr. Deutcher repeated three times that one should not blame your friends' academicians who simply believed what they were told.

(Solvay: Scientists should only agree on the basis of evidence if they think otherwise than what they are told).

I replied that it was not the first time that German scientists believed on simple assertion of superiors, and that some of them who know you well should have doubted.

(Solvay: The 93 intellectuals).

I said that they could have, with openness, made you appear in June, and that they would have noticed an erroneous interpretation or understanding of words. This would have avoided reprisals unjustifiable in view of your personality and your age. Wouldn't the learned world be surprised to know that such great figures acted with such pettiness towards an inventor, a scientist, an elderly man to whom their action could have caused great harm.

(Solvay: Quite right).

I have therefore been instructed to report the above two facts to you: your words and the so-called intervention of the Berlin academicians. I gave to Mr. Deutcher my opinion that, to provide him with more pleasant details for his mission, you would tell him, perhaps, in a letter what you would consider that the occupying authority should know about this incident.

(Solvay: I couldn't do that...).
Following Solvay's instructions, Lefébure informed Lorentz, who replied on January 7, 1919:

Dear Mr. Lefébure,

You will find attached your letter of August which you kindly sent me, and in which you exposed the reasons which made that Mr. Solvay was denied a passport for Switzerland. To tell you the truth, I was very surprised – at least with regard to those of them who attended the Brussels meetings – by the role that the Berlin physicists and chemists would have played. Knowing Mr. Solvay, they could be sure that he would not associate himself with the absolutely hostile feelings towards German science which have been expressed by French scholars; that on the contrary, he desires and hopes for the future an intimate collaboration of all nations.

It seems to me so obvious that they realized the true nature of Mr. Solvay's opinions that I am inclined to believe that, after all, they do not deserve our reproaches, and that Mr. Deutcher may have been misinformed. Of course, I am still talking about those who have been in contact with Mr. Solvay. Let us not forget that the German authorities, who unscrupulously deceived their people until the last moment, would not pay much attention to the advice of a scientist who wanted to deprive them of a pretext for the petty and unforgivable harassment they had in view.

In this regard, I believe I must tell you that in these last years, when I had to make some step in the interest of a Belgian, I often called for the help of Mr. Warburg, and especially of Mr. Planck, and that they have always lent me this help willingly and without any objection. I had to write several times to the German authorities, and I then asked Mr. Planck to send my letter where it could have the most effect. He has always done so, and has more than once associated himself with my request (...).

In cases wherever he didn't succeed, for example in that of the occupation of the Sociology Institute, Mr. Planck sent me the reply he had received and expressed his regret to me, speaking in terms that I believe were absolutely sincere. I owe him to tell you that. I could add other things, but I refrain from it, wishing not to waste your time...

Three days later, Lorentz addressed himself directly[516] to Solvay. He could no longer defer answering the many questions about ISIP's future (the renewal of its Scientific Committee, the attitude to be adopted towards German scientists...) and the industrialist's demands of August 14, 1916, concerning his project of social reform.

Dear Mr. Solvay,

Mr. Tassel sent me some time ago the enclosed letter which was sent to you by Mr. Haller. Mr. Warnant, through whom I received it, added that you would like to know what I think. He undoubtedly meant that, in connection with this letter, you asked me what I think of the future of the Physics Institute.

I hope you will excuse me for not responding to this more promptly. The question is important and my time has been very busy. However, I must not delay any longer. I therefore allow myself to tell you about my opinion on the main questions that seem to arise.

But I must tell you first that I was very sorry to learn that the events of the last two months had made you extremely tired, and that Mrs. Solvay suffered more than usual.

I hope with all my heart that you will both be better now and that you will fully enjoy the regained freedom. I read with extreme satisfaction that one of the first things for which the king took advantage of this freedom was to come and see you to thank you for all you have done, and that the government has bestowed on you the greatest honour it could.

I now come to the matters in question, but of course, you should not attach too much importance to what I will say about them. It is up to you and to the members of the Administrative Commission to take the necessary decisions.

1. *Will the Physics Institute still be able to do useful work by continuing the path that has been traced for it by the statutes, that is to say by encouraging scientific research in physics and in physical chemistry, and by organizing from time-to-time meetings for an in-depth discussion of major current issues?*

I do not hesitate to answer affirmatively. Now that the world will recover, the physical sciences will have to play – I'm sure – an increasingly important role. Mankind couldn't live without them, and would be failing in one of its earliest duties if it did not seek to penetrate the secrets of nature and to discover the truth above all.

I discussed all this with Mr. Kamerlingh Onnes and Mr. Verschaffelt. They entirely share my opinion. We are sure that, continuing as it began, the Institute will be able to prove itself worthy of your trust, and that it will be able to achieve, at least in part, what you had in mind founding it.

2. By contributing in this way to the progress of science, the Institute will have an indirect (underlined) *influence from a moral and social point of view. But will it be able to act more or less directly in this direction?*

That is a question which relates to what you wrote to me from Pontresina in August 1916.

First, I have to tell you that the action you then wished to see taken has hardly been started. I wrote to Mr. Guye and he gave me a provisional answer, but we stayed there. To go further, we should have listened to each other verbally, Mr. Guye, Mr. Haller and me, and that is what circumstances didn't allow. I therefore cannot tell you whether, perhaps, the three of us would have found a mode of action which would correspond to your ideas, while not straying too far from the goals for which the Institutes were founded.

I can only give you my personal opinion and, to tell you the truth, I am afraid that on this subject we do not agree. It comes from the very difference of what life has demanded of you and me. The problems which you had to solve have developed in you the great overviews. Your efforts have been aimed simultaneously at the progress of science and at social reform, or rather, these two objectives merged for you into one goal that you have pursued throughout your life with increasing zeal.

Me, on the contrary, I had to confine myself to my research and to my teaching. The special problems of physics demanded all my strength, and I can only be interested in questions of social or economic order without taking an active part in it.

I can admire what is done so that social development may lead, as far as possible, to the happiness and the well-being of mankind, but, for my part, I must simply remain a physicist for the years that I still have. You can claim the superiority of physical science; I must consider myself as a plowman who seeks to clear his land as others cultivate theirs, with the continuously revived feeling of the smallness of this piece of land and the greatness of the difficulties encountered. I have no doubt that you will understand my way of seeing, which, moreover, does not exclude the indirect influence that I just mentioned.

3. What should be our attitude towards the Germans?

The misery and the suffering they have spread over the world, the injustices and atrocities committed by their government and their armies, and justly abhorred by all honest people, have made on me – you know it well – a very strong and painful impression. Also, I understand perfectly that at this moment the Belgian and French scientists no longer wish to have any relation with them. For the first years to come there will be no place for them in the Physics Institute. Those who attended the Councils will understand that they cannot come to Brussels, and if there were others who would have the effrontery to do so, we would despise them precisely for that.

However, when speaking of the Germans, we must not lose sight of the fact that there are representatives of all the nuances. A man like Einstein, the great and profound physicist, is not "German" in the sense that the word is often used now. His judgment on the events of the past years will hardly differ from yours or mine.

There are also several physicists who refused to sign the scandalous Manifesto of the 93. In a conversation I had with them in the spring of 1915, Mr. Warburg strongly expressed his disapproval of the declaration. Certainly, a number of signatories would still do the same if the opportunity presented itself, but I am sure that many others deeply regret the lightness with which they have lent their name to this publication. You know the letter that Mr. Planck sent me and which I published at his request. I must also remind you of the question referred to in the appeal addressed to the 93 by Mr. Massart, a professor at your University in Brussels. He asked them to take the initiative of an impartial investigation into the facts of which the Belgian population was accused. This call went almost unanswered, like that of the Belgian bishops and the Freemasons. However, during a stay in our country, I spoke with Mr. Einstein about such an investigation into everything that happened in Belgium. He immediately recognized its absolute necessity and, after his return to Berlin, he discussed the matter with several members of the Academy of Sciences, in particular with Messrs. Waldeyer, Nernst and Planck. They told me shortly afterwards that they, too, wanted such an investigation to take place as soon as circumstances would permit, that is - as they thought - after the war. You know that now voices have been raised in Germany in favour of an impartial examination of the accusations in question and of the acts committed by the Germans. The scientists I just named could take part in it, but we must give them time. For the moment, the total collapse of their country and the danger of a Bolshevik revolution must occupy their minds entirely.

All things considered, I believe I should suggest to you not to exclude all Germans, that is to say, not to close the door to them forever. I hope that it can be opened to a new generation, and even that, perhaps, in the course of the years, we will be able to admit those scientists of our days of whom we can believe that they sincerely and honestly regret the events that took place. This is how German science will be able to resume its place which, despite everything, it deserves by its antecedents.

4. Last question. How should the Physics Institute be reconstituted, and how should it get back to work?

Of the nine members of the Scientific Committee, four should have left on November 1, 1916. The mandate of the other four would expire on November 1 of this year, while mine would last until 1922.

Since the partial renewal requested by the statutes for 1916 didn't take place, and since we do not know which of the members should have left, the composition of the Committee does no longer conform to the statutes.

> *In my opinion, I should therefore suggest to my colleagues on the Committee that we all offer our resignation to the Administrative Commission, or that we ask its members to consider us all as resigners. For myself, it seems correct to me not to take into account my exceptional position.*
>
> *Together with you, the Administrative Commission will then be able to choose a new Scientific Committee which it will compose in the manner it deems appropriate on account of the current circumstances.*
>
> *I ask you to communicate this letter to Messrs. Héger and Tassel. As soon as I know your, and their opinion on the questions I considered, I will be able to write to my colleagues on the Scientific Committee, and after that more or less officially address the Administrative Commission...*

There is no doubt that Solvay was disappointed by point 2 of Lorentz's letter, from which he deduced that he wouldn't obtain the desired social commitment from his scientific partners... However, he consoled himself with Lorentz's words about the future of the Physics Institute.

These are the words[517] he added to Tassel's response[518] of January 21, 1919, to Lorentz's letter:

> *"All my regards to you, my good Mr. Lorentz. I deeply reproach myself for having disturbed you beyond what I could have expected, and for having given you the trouble that I sense with regret. However, I thank you wholeheartedly for what you have done, and which will help in the grave circumstances we are going through. Tassel expresses my feelings well. I remain tired, but in good health. I should take a month's vacation. I hardly know how to write anymore...".*

8.7 Resumption of ISIP's activities

Lorentz's conciliatory attitude towards certain German physicists was by no means shared by Brillouin, who declared in a letter[519] to Solvay of November 24, 1918:

> *"Any international meeting requiring mutual esteem has become impossible with German scholars, whatever their intellectual merit...".*

Aware of the Frenchman's intransigence, Solvay instructed Tassel to send him the following message (letter[520] to Brillouin of January 21, 1919):

> *"Mr. Solvay is very convinced that if personal relations between scientists belonging to belligerent countries will remain impossible for a long time, scientific relations can never be suppressed. Such is also, I am sure, the feeling of our president, Mr. Lorentz, whom I know to have felt the misery and the sufferings widespread in the world by the Germans, as well as the injustices and atrocities committed by their government and their*

army, and who thinks as we do that there can be no place for them in the Physics Institute in the years to come...".

Marie Curie expressed her feelings in a letter[521] to Lorentz of February 26, 1919:

"May a brighter future erase the oppression and horrors of the past. I sincerely admire Mr. Solvay who, despite his age, fatigue and suffering has shown you his interest in resuming the activity of the Physics Institute. I completely agree with you as to the impossibility of admitting German scientists there at present. Their attitude about the invasion of Belgium would be enough, in my opinion, to make their presence in that country inadmissible as guests of Mr. Solvay. Moreover, you would not find, I think, French or English scholars willing to resume relations with those who, forming the elite of their country, did not encourage a high moral and humanitarian ideal but the worst spirit of brutal reaction.

I believe that the German teaching staff, and at its head the university professors, are responsible for the state of mind which in Germany has led to unleash the war.

As far as I am concerned, I authorize you to offer the Administrative Commission the resignation of the Scientific Committee. This solution seems to me to be entirely in accordance with the circumstances. It is desirable that the Commission should be able to reorganize the Committee in complete freedom...".

At about the same time, Warburg and Nernst informed[522] Lorentz that they accepted the idea of a collective resignation of the ISC members. Rutherford did the same in a letter of March 5, 1919, indicating that it also seemed difficult to him to resume, for a certain time, contacts with his German colleagues.

On May 12, Tassel announced to Lorentz that the Administrative Commission had accepted the collective resignation of the members of the Scientific Committee.

The activity of the Institute having been suspended during the years of war, it seemed natural to the Commission to consider these years as non-existent. Tassel proposed to use the corresponding annuities so as to extend ISIP's lifetime beyond what had been foreseen, while increasing its annual income. He suggested an increase to 60 000 francs, payable on July 1 of each year from 1919 until 1948, so as to restore the thirty years provided for the duration of the Institute at the time of its foundation.

Tassel also recalled the need to modify the Institute's statutes with regard to the duration of the mandate of the Committee's president. As to the renewal of this Committee, it seemed desirable to replace Nernst and Warburg, by new members. The names put forward were those of the British physicist William H. Bragg, one of the rapporteurs to the second Solvay Council, and of the Italian senator Augusto Righi, a specialist in electromagnetic waves.

Goldschmidt's bitterness

In addition to the above measures, it seemed appropriate to ask Goldschmidt to join ISIC's newly formed Scientific Committee, and to cede his position in ISIP to the Belgian physicist Edmond van Aubel. The proposal was very badly received by the Councils' "general manager", who blamed Lorentz[523] for having disregarded the many services rendered to the Physics Institute, and for having underestimated his merits as a physicist: creation of a laboratory for the study of electromagnetic waves, setting up of wireless telegraphy installations, participation[524] in the "Commission Internationale de TSF Scientifique"... But Goldschmidt's main reason for bitterness was the fact that his departure from ISIP came at the time of the German members' exclusion, and that it would be attributed to his German background, or to his former scientific connection with Nernst.

Publication of the report of "Solvay II"

ISIP's Scientific Committee having been reorganized, it became imperative to address a delicate question: the publication of the official report of the second Solvay Council.

Tassel raised the question in a letter[525] to Brillouin of June 13, 1919:

"You may recall that at the urging of the German Committee members it had been decided – even though the idea testified to a feeling of little respect towards the founder – that the text of the minutes would be printed using the languages in which the various communications had been made (...). We embarked without the slightest enthusiasm upon the printing of this baroque multilingual assembly.

Today it seems to us impossible to publish the minutes in this form, after the events which have taken place. The Commission therefore proposed to Mr. Lorentz to have the English and German parts translated, and to publish the report in French.

It will be done as follows: Mr. Verschaffelt has accepted to take care of the translations (Mr. Lorentz will ask Mr. v. Laue in Zurich the necessary authorization regarding his report)... We took advantage of the revision of the statutes to specify that the French language will henceforth be adopted for the publication of the minutes of the Physics Councils. This decision is in accordance with the founder's desire.

Needless to say that this measure will not exclude the presentation of reports drawn up in a foreign language, nor the use of these languages in discussions (...). Nothing was published in relation to Mr. Solvay's foundations during the war. I carefully avoided during the occupation any action or manifestation that could draw the attention of the occupying power to our international Institutes. I feared their confiscation...".

In compliance with the Commission's decision, the publication of the minutes of the second Physics Council was entrusted to Gauthier-Villars. The volume, entitled "*The Structure of Matter*" appeared in 1921.

As to ISIP's future activities, it was decided that they would be discussed at the next meetings of the renewed Scientific Committee, scheduled for March 30 and 31, 1920.

Report of the first post-war ISC meetings[526]

1. General discussion

Regarding ISIP's new priorities, Lorentz explained that the depreciation of the Belgian franc[527] and the increase in the cost of instruments made it impossible for the Institute to continue subsidizing research projects in various countries. However, in order to comply with Solvay's wishes, he proposed to set up a financial reserve to be used in certain cases of exceptional importance.

The granting of regular subsidies being no longer on the agenda, it was decided that ISIP's main task would be the periodic organization of "Physics Councils". These meetings would take place every three years. They would involve a maximum of twenty-five members (in particular for reasons of economy). Rutherford and Marie Curie welcomed the decision, indicating that a small number of participants was an essential factor for the Council's success.

2. Organization of the next Council

The third Solvay Council on Physics was scheduled for the beginning of April 1921. Regarding the issues to be discussed, the president announced that W. H. Bragg and A. Righi (who couldn't be present at the meeting) had expressed their opinion by letter. These were the ideas put forward:
- Righi proposed to discuss the experimental foundations of the theory of relativity; he wished the presence of Tullio Levi-Civita (in fact, we know[528] thanks to a personal letter to Solvay, that Righi intended to discredit Einstein's theory...).
- Bragg didn't favour a discussion of relativity theory. He wanted to devote the Council to the structure of molecules, atoms and nuclei, and suggested to discuss the origin of magnetism.
- Marie Curie and Brillouin pointed out that the theory of relativity didn't raise pressing questions, and that there were a number of open problems in molecular theory, the discussion of which should lead to immediate work. Marie Curie also raised the question of the application of quantum theory, insisting on the existence of a gulf between theory and logic.
- Lorentz explained that a discussion of relativity theory could wait, and that the only question to be examined was the displacement of spectral lines in the theory of gravitation.

- Kamerlingh Onnes pointed out that the problem of relativity was too abstract to be dealt with in a physics Council. He suggested to devote the next meeting to magnetism, a subject which had been hardly studied by the Germans and which could therefore be discussed in their absence.
- Rutherford proposed to focus on the theory of electrons and on their role in a variety of domains, ranging from magnetism to radiation. Marie Curie supported the idea, indicating that it would constitute a coherent whole.
- Lorentz closed the discussion, favouring the subjects suggested by Rutherford and Marie Curie[529]. He proposed to concentrate the next Council on the role played by the electrons.

Conclusion: the Committee decided that the Council, entitled "*Electrons, atoms and radiation*", would involve the discussion of 4 reports, with a total of 6 presentations:

1. *Classical theory of electrons, application to the theory of radiation (emission, absorption), scope and limits of the theory* (Lorentz).
2. a. *Structure of the atom, charge and constitution of the nucleus, isotopes* (Rutherford).
   b. *Intervention of quantum theory, Einstein's relation* (M. de Broglie).
3. *Bohr's spectral theory, methods for understanding the arrangement of electrons in the atom* (Bohr).
4. a. *Electrons and magnetism, gyroscopic effects* (Einstein, Richardson or de Haas).
   b. *Attempts to explain para- and diamagnetism, superconductivity* (Kamerlingh Onnes or Langevin).

The Committee also drew up a list of participants:
- United Kingdom: Rutherford, W. H. Bragg, O. Richardson, Jeans, J. J. Thomson, Barkla, Larmor.
- France: Marie Curie, Brillouin, Langevin, M. de Broglie, Perrin, Weiss.
- Netherlands: Kamerlingh Onnes, Lorentz, W. de Haas, P. Zeeman, Ehrenfest.
- Scandinavia: Bohr, Knudsen, Vegard or Siegbahn.
- Italy: Righi.
- Switzerland: Einstein[530].
- United States: Millikan.
- Belgium: van Aubel.

In reserve: J. C. Mc. Lennan, Ch. Fabry, A. Cotton, W. H. Keesom, L. Brillouin, W. L. Bragg, M. Barnett, A. Nagaoka, E. Henriot.

Lorentz indicated that he would insist on the presentation of brief reports (not more than 1 hour) designed to shed light on the difficulties of each subject dealt with.

Marie Curie suggested that some time for reflection be given after the discussion of each report, to allow the drawing up of a summary of the ideas put forward, and the formulation of the Council's conclusion for each issue.

Temporary exclusion of German and Austrian physicists

The First World War could have signalled the end of ISIP. The danger of a dissolution was real, in particular because of Brillouin's intransigence. Yet, the Institute survived thanks to the conjunction of three factors: Lorentz's positive and conciliatory attitude, the understanding shown by Solvay, the prudence of Héger.

ISIP's statutory provisions meant that no member of the Scientific Committee was formally excluded. All mandates having expired, except that of the president, the members were simply invited to present their collective resignation. Hence, the German Committee members didn't have valid reasons to feel humiliated.

Contrary to the exclusions pronounced by the sections of the International Research Council[531] (created by the Allies to replace the International Association of Academies), ISIP's authorities took care not to permanently close the door to physicists from the Central Powers. This wisdom enabled Solvay's Institute to survive the collapse of international relations.

We shall see (section 9.3) that in April 1926 the Scientific Committee decided to invite four German scientists to attend the fifth Physics Council, thus restoring the Institute's universal mission. The Committee's decision to put an end to ISIP's subsidy program (a program of which the Germans had been the main beneficiaries), was in no way dictated by an anti-German sentiment, it was the consequence of a concrete situation: the devaluation of the Belgian franc.

Chapter 9
Epilogue: from "Solvay III" to "Solvay V"

It would take us too far to evoke the first Chemistry Council, held in April 1922, and the numerous Physics Councils which followed one another from the Great War to our days. However, it seemed legitimate to end our story with an overview of the last three Councils chaired by Lorentz.

We will focus on Solvay V, the Council which in 1927 welcomed again some German physicists, and which marked the advent of a new era in physics with the birth of quantum mechanics. It will allow us to close the loop, opened sixteen years earlier with the first Solvay meeting.

9.1 Solvay III: Atoms and Electrons, April 1–6, 1921

The program of this first post-war Council was determined, as reported in section 8.7, on March 31, 1920.

Participants to the meeting were accommodated at Hotel Britannique, located on Place du Trône, near the gardens of the Royal Palace (the building still exists today). The Council sessions were held at Parc Léopold's Physiology Institute (figure 41).

Extract from the conference report[532]:
Chairman: H. A. Lorentz.
Participants:
i. Members of the ICS: Marie Curie, H. Kamerlingh Onnes, M. Brillouin, P. Langevin[533], E. Rutherford, E. van Aubel, M. Knudsen (secretary).
ii. Invited members: C. G. Barkla, N. Bohr, W. L. Bragg, L. Brillouin, M. de Broglie, W. J. de Haas, P. Ehrenfest, J. H. Jeans, J. Larmor, A. A. Michelson, R. A. Millikan, J. Perrin, O. W. Richardson, M. Siegbahn, P. Weiss, P. Zeeman.

FIGURE 41: Members of the third Solvay Council on Physics (1921), Brussels, Solvay Physiology Institute, Parc Léopold, S.a.b.ULB. Courtesy of the International Solvay Institutes for Physics and Chemistry.

FIGURE 42: William Lawrence Bragg in 1921, see figure 41.

Reports

1. *The classical electron theory (scope and limit)*, by H. A. Lorentz.
2. a) *Structure of the atom (charge and constitution of the nucleus, isotopes)* by E. Rutherford.
   b) *Intervention of quantum theory (Einstein's relation in photoelectric phenomena)* by M. de Broglie.
3. *Bohr's spectral theory (distribution of electrons in the atom)*, by N. Bohr.
4. a) *Electrons and magnetism (gyroscopic effect)* by A. Einstein or O. W. Richardson.
   b) *Attempts to explain para- and diamagnetism; magnetism at very low temperatures*, by W. de Haas.

Some additional papers were presented to the Council, in particular by Kamerlingh Onnes on *Magnetism at very low temperatures*, by L. Brillouin on *The conductivity of metals*, by R. Millikan on *Quantum absorption of radiation in metals*, and by P. Weiss on *The mutual action of magnetized molecules*.

Comments

Einstein and Bohr, the two most anticipated guests, weren't present.

Einstein, regarded as a Swiss invitee, was prevented from going to Brussels because of a trip[534] to the United States. However, he took part in the Council through an intermediary: his contribution (the theoretical part of report 4b) was presented by Wander de Haas, Lorentz's son-in-law.

Bohr was exhausted by the founding of the Copenhagen Institute of Theoretical Physics. He also participated in the Council through an intermediary: his report was presented by Ehrenfest.

The Council welcomed two representatives of a new generation: William Lawrence Bragg (figure 42), son of W. H. Bragg[535], and Léon Brillouin (figure 43), son of M. Brillouin.

Were also present: two famous American physicists, Albert A. Michelson[536] (figure 44), Nobel laureate in 1907 for his work in optics, and Robert Millikan (figure 45), Nobel Prize in 1923 for his research on the elementary unit of electricity, and his work on the photoelectric effect.

Lindemann wished to attend the Council as an auditor[537]. He had made this clear to Lorentz in January 1921, reminding him of the services he had rendered as scientific secretary of two Councils[538] (Lindemann knew that auditors weren't admitted). Lorentz felt embarrassed but persisted in his refusal. Anxious to preserve the confidentially of the debates, he told Lindemann that M. Brillouin

FIGURE 43: Léon Brillouin in 1921 (standing, centre of the picture), see figure 41.

FIGURE 44: Albert A. Michelson in 1921 (sitting, centre of the picture), see figure 41.

FIGURE 45: Robert A. Millikan in 1921 (sitting, centre of the picture), see figure 41.

FIGURE 46: Maurice de Broglie in 1921 (right hand side of the picture), see figure 41.

would have strongly objected against the presence of a man who had worked in Berlin, and who had been one of Nernst's closest collaborators[539].

The Council's discussions focused on two fundamental questions.

One was related to matter. It concerned the structure of the atom: replacement of Thomson's model by the nuclear model of Rutherford-Bohr.

The other question concerned the "dual" nature of light: arguments by Bragg and by Barkla in favour of the classical wave concept, *versus* M. de Broglie's analysis of *pin-shaped* X-rays, suggesting a particle-like structure of radiation[540].

In his brilliant report, de Broglie (figure 46) enumerated a series of phenomena in which quanta seemed to intervene in an individual way: photoelectric effects and inverse phenomena. His results validated Einstein's relation between the energy of a light quantum and the frequency of the radiation. They also provided decisive support to Bohr's bold postulates.

Two particular events caught the members' attention:
- Rutherford's prediction of the neutron (its existence would be confirmed in 1932 by one of Rutherford's pupils, James Chadwick).
- The convincing answer that de Broglie's spectacular study of X-rays had brought to M. Brillouin's remark of 1911 (general discussion[541] of the first Solvay Council): "*It isn't very satisfactory to be reduced to recognizing the presence of a discontinuity by means of apparently continuous phenomena, to introduce it at the basis of a theory, before drowning it out with the help of statistical considerations. If we could imagine an experiment which made us seize the discontinuity "on the spot", it would be much more decisive and instructive...*".

FIGURE 47: Manne Siegbahn in 1921 (standing, centre of the picture), see figure 41.

De Broglie's presentation of his X-ray experiments made a great impression, in particular on Manne Siegbahn (figure 47), future Nobel laureate[542] in physics (1924), and first Swede to be invited to a Solvay Council. Siegbahn shared his enthusiasm with his compatriot Carl Oseen, who became a member of the Nobel Physics Committee in 1922. This nomination would allow him to work for the attribution of a Prize to his friend Bohr. But Oseen was confronted to a problem: how could the Nobel Committee honour Bohr before awarding a Prize to Einstein?

A solution was easily found: the 1921 Prize hadn't been awarded. This enabled Oseen to become the 'architect" of two historic Prizes[543]: that of 1921, awarded to Einstein in 1922 *"for having established a link between quanta and the photoelectric effect"*, and that of 1922, awarded to Bohr *"for having discovered the link between quanta and the structure of the atom"*.

9.2 Solvay IV: Electric Conductivity in Metals and Related Problems, April 24–28, 1924

Lorentz also chaired Solvay IV, the first Council which took place after Solvay's death. He addressed a telegram to Ernest's widow on behalf of all members.

Adèle Solvay answered[544] on April 2, 1924:

"It is a deep satisfaction for me to see the work of my husband continued thanks to the devoted zeal of the new Physics Council. It is with emotion that I address you, and the members of the Council, the expression of my sincere gratitude for your telegram full of sympathy."

The members were invited to dinner[545] by her son, Armand Solvay, on April 28, 1924. Some guests, in particular Marie Curie, were accommodated at the recently built hotel of the "University Foundation", rue d'Egmont (the hotel still exists today). Others stayed in Grand Hôtel Britannique.

The Council sessions took place, as before, at Parc Léopold's Physiology Institute (figure 48).

Extract from the conference report[546]

Chairman: H. A. Lorentz.

Participants:
i. Members of the ICS[547]: Marie Curie, H. Kamerlingh Onnes, M. Brillouin, W. H. Bragg, E. Rutherford, P. Langevin, E. van Aubel, M. Knudsen (secretary).
ii. Invited members: H. Bauer, P. W. Bridgman, L. Brillouin, W. Broniewski, P. Debije, E. H. Hall, G. de Hevesy, A. Joffé, W. H. Keesom, F. A. Lindemann, O. W. Richardson, W. Rosenhain, E. Schrödinger.

FIGURE 48: Members of the fourth Solvay Council on Physics (1924), Brussels, Solvay Physiology Institute, Parc Léopold, S.a.b.ULB. Courtesy of the International Solvay Institutes for Physics and Chemistry.

Reports

1. *Application of the electron theory to metals and their properties* (Lorentz).
2. *Report on conductivity phenomena in metals and on their theoretical explanation* (Bridgman).
3. *A theory of conductivity in metals* (Richardson).
4. *The internal structure of alloys* (Rosenhain).
5. *Electrical resistance and metal expansion* (Broniewski).
6. *Electrical conductivity of crystals* (Joffé).
7. *New experiments with superconductors* (Keesom[548]).
8. *Metal conductivity and the transverse effects of the magnetic field* (Hall).
9. *Note on the propagation of radiation impulses* (Joffé and Dobronravoff).

Comments

Lorentz, as we saw, had taken care to devote Solvay IV to a question that could be dealt with in the absence of German physicists. The subject "*Electrical Conductivity in Metals*", related to his famous electron theory, probably came to his mind during a stay in the United States.

Indeed, we know that Lorentz made a transatlantic tour from January to April 1922, allowing him to meet several American and Canadian physicists. These included three invitees to Solvay IV[549] and one to Solvay V: Edwin Hall, the discoverer of the phenomenon that bears his name; Percy Bridgman, a specialist in high pressures (Nobel Prize in 1946); Arthur H. Compton, famous for his discovery of a phenomenon which confirmed the existence of the photon (invitee Solvay V and Nobel Prize in 1927); John MacLennan, the low-temperature expert in Toronto, who in 1923 would produce liquid helium, the first physicist to do so after Kamerlingh Onnes in 1908.

Lorentz lectured during his trip in eleven physics centers, notably in Pasadena (where he stayed from January 5 to March 3), in Madison, in Toronto, and finally on the East Coast and in Washington.

In Madison, where he spent a week, Lorentz participated in a symposium. He played the part he used to play in Solvay Councils: presentation of a paper and discussion of reports presented by colleagues. Two speakers were physicists whom he would meet again in the US: Compton who was about to discover his famous effect in Chicago, and Bridgman, a researcher attached to Harvard University.

FIGURE 49: Erwin Schrödinger in 1924 (standing in the back of the picture), see figure 48.

During his visit of Harvard, Lorentz stayed with Hall. In Toronto he stayed with MacLennan, an invitee to Solvay IV who finally announced that he wouldn't come.

Einstein's attitude came as a surprise. He declined the invitation to Brussels, refusing to be once more the only German guest, and begged ISIP not to invite him again.

On the other hand, the Council welcomed Erwin Schrödinger (figure 49), an Austrian physicist (Nobel Prize in 1933), attached to the University of Zurich, and George de Hevesy (figure 50), a Hungarian researcher (Nobel Prize in 1943 for his pioneering work

FIGURE 50: George de Hevesy in 1924 (sitting, centre of the picture), see figure 48.

on radioactive tracers) who worked in Copenhagen. It is probable that, as he launched the invitation, Lorentz was aware that it would mitigate the impression of a permanent boycott of scientists from the Central Powers.

However, the presence of Schrödinger and de Hevesy at Solvay IV is remarkable for another reason. It is indicative of the insight of the members of ISIP's Scientific Committee, who didn't hesitate to invite promising physicists who had not yet made the discovery[550] that would make them famous.

Despite the absence of German physicists (including Einstein), Solvay IV should be seen as a Council "open to the world". It counted among its members two American physicists: (Bridgman and Hall), and a Russian rapporteur: Abram Fedorovich Joffé (figure 51), considered today as the father[551] of Soviet physics.

FIGURE 51: Abram F. Joffé in 1924, see figure 48.

Anecdotical event: a Brussels newspaper published an interview with Marie Curie, recorded on the meeting's sidelines. She was asked to comment on an announcement that had caused a sensation: the alleged discovery of a new "invisible and diabolic" radiance. The event is revealing of a growing awareness which manifested itself in 1924: the dangers[552] of the radiations emitted by radium.

9.3 Solvay V: Electrons and Photons, October 24–29, 1927

The members of the ISC met in Brussels[553] during the first days of April 1926. They fixed the program of the fifth Physics Council.

Kamerlingh Onnes had died a few weeks earlier. Langevin insisted that his seat be allocated to Einstein[554]. The proposal was unanimously accepted.

Impressed by the recent German results in quantum theory, Langevin and other Committee members agreed with Lorentz that it was time to invite "moderate" German researchers. These proposals – Einstein's election as a member of the ISC and the invitation of German physicists – were submitted for approval to ISIP's Administrative Commission. Their acceptance was facilitated by the Locarno Agreements of October 1925, and by Germany's decision to join the League of Nations[555].

Yet, the presence in Brussels of German physicists remained a sensitive question... It required the king's approval.

The meeting of the Scientific Committee having been fixed for March 21, 1926, King Albert had indicated that he would receive the members for lunch. He also announced his wish to have a private meeting[556] with Chairman Lorentz. The audience took place on April 2. Einstein's nomination was approved. After having remarked that seven years after the war the feelings it provoked should gradually soften, the king expressed the opinion that "*a better understanding between the peoples was absolutely necessary for the future, and that science should help to bring it about*". His final words were in perfect agreement with Lorentz's feelings: "*Considering what the Germans have done in physics, it would be difficult to do without them.*"

The Committee's proposals

It was decided that the issue of the fifth Physics Council would be "Quantum and classical theories of radiation".

A provisional list of guests comprised fifteen names[557]: Bohr (Denmark); Kramers, Ehrenfest (Netherlands), Fowler, W. L. Bragg, C. T. R. Wilson (United Kingdom); L. de Broglie, L. Brillouin[558], H. A. Deslandres (France).

A. H. Compton, (United States); E. Schrödinger (Austria[559]); P. Debije (Switzerland); Planck and two physicists to be chosen from M. Born, W. Heisenberg and W. Pauli (Germany).

Substitutes were: M. de Broglie or Thibaut for Bragg; Dirac for Brillouin; Kapitza for Wilson; Fabry for Deslandres; Darwin or Dirac for Fowler; Bergen Davis for Compton; Thirring for Schrödinger.

Seven reports were on the program:
1. *New verifications of classical radiation theory* (W L. Bragg).
2. *The Compton effect and its consequences* (Compton or Debije).
3. *Observation of photoelectrons and shock electrons by the condensation method* (Wilson).
4. *Interferences and light quanta* (L. de Broglie).
5. *A note by Mr. Kramers on Slater theory*, if deemed useful (Kramers).
6. *New derivations of Planck's law; applications of statistics to quantum theory* (Einstein or Ehrenfest).
7. *Adaptation of the foundations of dynamics to quantum theory* (Heisenberg or Schrödinger).

Comments

Despite the success of the Bohr-Sommerfeld theory, no report was to discuss the role of quanta in atomic physics (the focus was on the application of quantum theory to radiation). Yet, it should be noted that Bohr's name was on top of the list.

That Sommerfeld wasn't invited may seem surprising[560] since we know that Lorentz held him in high esteem[561], and that he congratulated him in November 1924 for his "*new great services to science*[562]". Also Born was surprised not to see Sommerfeld's name on the list of invitees. He wrote him a letter[563] on June 15, 1926, to express his astonishment.

To understand ISIP's attitude with regard to Sommerfeld we must go back to the year 1918 and to the latter's actions in Belgium during the German occupation (an issue already discussed in section 8.3). How could the members of ISIP's Administrative Commission accept the presence of a man who supported Flemish activists, and who had been happy to teach German soldiers in a place which, in all probability, was part of an occupied Solvay Institute?

In short, the Munich professor had made himself undesirable in Brussels.

Yet, Sommerfeld's exclusion wouldn't last forever. In 1930 he was invited to take part in Solvay VI, a meeting chaired by Langevin, and devoted to magnetism. We may assume that Lorentz's successor – a Frenchman close to Einstein and to Auguste Piccard – intervened with all his weight to get Sommerfeld invited to the first Council that would take place under his presidency.

Back to Solvay V and to its preparation[564]

Important steps took place in April 1927. Barely returned from the United States, Lorentz announced to the Secretary of the Administrative Commission (Ch. Lefébure) that all invitees had answered the call (with the exception of astronomer Deslandres), and that he would ask the rapporteurs to get down to business[565].

Frustrated by the thirty-six months that had elapsed between the holding of Solvay IV and the publication of the report, Lorentz decided to take action. He told Lefébure:

"The method we have followed so far doesn't appear to be the right one. I propose to write, with Verschaffelt's help, the Council discussions in the first two months, without the original notes having been sent to the members (as we have done so far). They would only receive the completed proofs, and we would ask them to return them before a fixed date...".

Three months later, Lorentz was informed of the latest developments:
- Einstein, who had been asked to talk about his work on perfect gases (application of a new derivation of Planck's law, proposed by the Indian physicist Satyendra N. Bose), made clear on June 17 that he didn't wish to present a report[566].
- W. H. Bragg, another member of the ISC, announced on August 27 that he wished to retire from the Committee and that he wouldn't come to Brussels.
- E. van Aubel, the Belgian Committee member, explained that he refused to take part in a Council attended by German physicists.

Following these defections, Lorentz decided to launch new invitations. Relying on Ehrenfest's advice[567], he sent this message to Lefébure[568]:

"Since last year, quantum mechanics, which will be our subject, has developed with unexpected rapidity, and physicists, who first came in second

place, have made extremely remarkable contributions. That's why I would be happy to invite Mr. Dirac (from Cambridge) and Mr. Pauli (from Copenhagen)."

The proposal would prove crucial. The Council would benefit from the presence of Born, Heisenberg, Pauli (figure 52) and Dirac (figure 53). Members would be duly informed of the most recent advances in quantum mechanics.

Yet, Lorentz had still other problems to tackle. The invitation of German scholars continued to cause a local stir. Members of the Administrative Commission expressed fear of manifestations[569]. The Dutchman had no choice but to go to Brussels to discuss the matter with Lefébure. The meeting took place on October 17. It was decided[570] that the list of all invitees wouldn't be communicated to the press.

Lorentz also had to refuse a last-minute request from Ch. Manneback, a professor at the University of Louvain, whom he had met in Como[571] in September 1927, and who wished to attend the Council as an observer. Lorentz stood firm: he reacted negatively, as he had done in 1921 with Lindemann. On the other hand, he welcomed three professors from the Université Libre de Bruxelles, A. Piccard, Théophile de Donder, and Jacques Errera, declaring[572]:

> *"It would be shocking for them to meet us at the university reception and at Mr. Solvay's dinner, and not be invited to our sessions. In addition, we now have relations with the Free University that we do not have with other universities in the country."*

But that wasn't all. Lorentz's most remarkable initiative was a proposal to invite Irving Langmuir (figure 54), an American chemist attached to General Electric.

The man was staying in Europe to take vacation with his wife. Having broken his foot during the trip, he was taking a rest in Italy...

FIGURE 52: Max Born, Werner Heisenberg, Wolfgang Pauli in 1927: Born (second rank, right hand side of the picture), Pauli (third rank, centre of the picture), Heisenberg (third rank, right hand side of the picture), see figure 57.

FIGURE 53: Paul Adrien Maurice Dirac in 1927 (second rank, centre of the picture), see figure 57.

FIGURE 54: Irving Langmuir in 1927 (sitting, centre of the picture), see figure 57.

Here is the telegram[573] that Lorentz sent to Lefébure on October 19: *"Langmuir, excellent American physicist now in Paris. May I invite him to the Council?"*.

The message is intriguing. Langmuir was equally intrigued by the invitation. His astonishment appears in the few words he sent to his mother[574]:

"This year, the Council focuses on quantum theory, an area to which I have not contributed at all. So, I do not see why I am invited...".

A key to understand Lorentz's decision can be found in *"Cathedrals of Science. The Personalities and Rivalries that Made Modern Chemistry"* by P. Coffey, a specialist in the actions of Gilbert N. Lewis and of Irving Langmuir. These are his words[575] about Langmuir's invitation to Solvay V:

"Léon Brillouin, the French physicist and an invitee to the conference had been in contact with Langmuir's boss, Willis Whitney, trying to negotiate a consulting contract for himself with GE, and Whitney had asked Brillouin to connect with Langmuir at an earlier conference in Como, so he apparently fudged things a bit to get Langmuir an invitation to the Solvay Conference. He led Lorentz to believe that Langmuir was going to be in Brussels anyway that week and asked him to extend Langmuir an invitation. Langmuir jumped at the chance and later sent Whitney a report of the conference:

"We had planned to leave Cortina on October 23rd and go to Eindhoven (presumably to visit the Philips Research Laboratory), but, on October 19th, I got a telegram from Prof. H. A. Lorentz (Haarlem) inviting me to attend the Solvay Congress which was to meet in Brussels for 1 week, beginning Oct.23rd, to discuss the quantum theory. Léon Brillouin had written to Lorentz suggesting that I be invited. I had heard about the Solvay Congress in Como in Sept. and had inquired about it even then from Lorentz, but was told that only those invited specially could attend, and the membership of about 25 was limited to the most active workers in the quantum theory – so I was surprised to receive an invitation...".

There still are two open questions. Why did Lorentz accede to Léon Brillouin's request, despite his well-known concern for strictly enforcing the Council's rules? And why did he qualify Langmuir as an *"excellent American physicist"*?

It is tempting to associate Lorentz's motives, and the terms of his telegram, with his recent stay in the United-States.

Let us go back to the end of 1926, when Lorentz visited Cornell and addressed the members of the American Chemical Society. Talking about quantum theory and the structure of atoms, he mentioned Langmuir's paper "*The Arrangements of Electrons in Atoms and Molecules*", published in 1919. This fact alone might explain Lorentz's qualification of Langmuir as a physicist... Another element is the fact that Langmuir had no difficulty whatsoever in grasping the scope and the significance of the latest advances in quantum theory. This is attested by the acuity with which he summed up the Council's main conclusions in his report of the Brussels debates[576]:

> "*Strenuous meetings – 8 hours per day in lecture room and all evening (often till 1 A. M.) in hotel where we all stayed, discussing quantum theory. The wave mechanics is the universal mode of expression today. For the first time the quantum theory can be formulated in apparently complete form, so that the discrepancy between the classical wave theory and the quantum phenomena (and even the photon theory) seems to be clearing up...*".

Last, but not least, we know that Langmuir had close relationships with leading physicists. He and his wife continued their European tour after the Solvay meeting and stayed with Bohr in Copenhagen[577]. They also visited Lorentz in Holland...

Yet, we still need to understand the more personal motives which led to Lorentz's receptive attitude towards Léon Brillouin. These were apparently linked to an event which will be discussed in the next paragraph.

However, before doing so, we should say that Brillouin's suggestion to invite Langmuir was a brilliant one... Indeed, the American had a camera and used it to make an exceptional document: a short film[578] in which we see the fathers of quantum theory walking on Brussels' Grand-Place, Bohr, Einstein, Ehrenfest, de Broglie, Schrödinger, Kramers, Born, Pauli, Heisenberg, Dirac and Debije.

Celebration of Fresnel's centenary

A peculiarity of Solvay V was that its sessions were interrupted on October 27. This was due to an event which had no relation with the Council: the celebration in Paris of the centenary of the death of Augustin Fresnel, a founding father of the wave theory of light.

On September 12, 1927, Lorentz learned that the Fresnel celebration would take place at the Sorbonne on October 27, a date chosen by the President of the Republic, and that the French Physical Society, organizer of the event, wanted to invite the members of Solvay V.

FIGURE 55: Pieter Zeeman in 1921 (sitting, left hand side of the picture), see figure 41.

FIGURE 56: Louis de Broglie in 1917 (centre of the picture), see figure 57.

One of the reasons of the invitation was that the organizers counted on a speech by Lorentz, and on lectures by P. Zeeman (figure 55) and L. de Broglie (figure 56).

The news was most embarrassing as the ceremony fell in the middle of the Brussels meeting. Moreover, it was to be expected that some members would wish to attend. Fortunately, Lorentz thought of a solution: he proposed to suspend the sessions programmed on the afternoon of October 27 and on the next morning, and to prolong the Council's debates on Saturday 29. To implement the idea, he needed some help. On October 10, Lorentz asked Brillouin[579] to take Langevin's advice. In case of the latter's agreement, Brillouin would be in charge of the necessary arrangements with the French Physical Society.

Langevin approved the idea. Brillouin made contact with Jean Thibaud, the Secretary of the French Physical Society. Two days later, Lorentz received official invitations from the Society's President, Louis Lumière[580]. Brillouin sent a list of Council members wishing to attend the ceremony. It comprised the following names: Bohr, Born, Bragg, Compton, Debije, de Broglie, de Donder, Einstein, Fowler, Guye, Heisenberg, Kramers, Langevin, Lorentz, Pauli, Richardson, Verschaffelt, Wilson.

The presence of German physicists at the Sorbonne, even in a private capacity, was a *premiere* after the war (Thibaud had indicated that German Societies had been invited, but that none had responded[581]). This fact was close to Langevin's heart as reflected three months later in this eulogy at Lorentz's funeral[582]. Addressing the deceased, he declared:

> "*It was your last visit to Paris, but a particularly precious one, since you had the touching thought of bringing with you the members of the fifth Solvay Council, first act since the war and happy omen of the resumption of a genuine and long-awaited international collaboration of all physicists.*"

A financial detail about the Paris event is revealing of the generosity of the Solvay family: the costs of the trip were covered by Ernest's widow, Adèle, who chose to remain discrete (she begged Leféb ire to tell the members that all costs had been paid by the Institute[583]).

Specificity of the fifth Solvay Council on Physics

The Brussels newspaper "Le Soir" of October 23, 1927, published an instructive note by Lorentz on the purpose and challenges of Solvay V:

"The issues dealt with by the present Council are closely linked to those of sixteen years ago. Mr. Planck had introduced the notion of discontinuities and abrupt transitions in the motion of atoms and electrons, and in 1911 the question was to appreciate the role that these discontinuities were playing in various phenomena, and to examine the fundamental laws that govern them.

This first exploration of a new and vast terrain, the richness of which has been revealed during the years that have passed since then, already suggested the need to reform the foundations of mechanics, so as to reveal the discontinuities in question (their magnitude is now referred to as "quanta") and to characterize them, not as something additional but as a fundamental and essential element.

This year the discussions will focus on the attempts that have been made in recent years to develop what may be called "quantum mechanics", and in which Messrs. de Broglie, Heisenberg, Born, Schrödinger, Dirac and others, took part. The work of these physicists still constitutes an assembly of a more or less disparate aspect, because in spite of the unity which exists deep inside, great divergencies of opinion remain.

As a result, the "clash of opinions", capable of bringing us closer to the truth, will not be lacking... It is in such circumstances that the Solvay method is particularly applicable, and that it has the capacity to clarify ideas and to accelerate progress."

Lorentz was right. Solvay V intervened at a key point in the development of quantum theory. Many advances had been made in recent months, ensuring heated debates.

As expected by the chairman, the Council members witnessed the emergence of new and revolutionary concepts, capable of providing answers to the many questions that had haunted researchers for more than fifteen years.

Extract from the conference report[584]

The Council sessions took place at Parc Léopold's Physiology Institute (figure 57).

FIGURE 57: Members of the fifth Solvay Council on Physics (1927), Brussels, Solvay Physiology Institute, Parc Léopold, S.a.b.ULB. Courtesy of the International Solvay Institutes for Physics and Chemistry.

Chairman: H. A. Lorentz.
Members: Mme Curie, N. Bohr, M. Born, W. L. Bragg, L. Brillouin, A. H. Compton, P. Debije, L. de Broglie, P. A. M. Dirac, P. Ehrenfest, A. Einstein, R. H. Fowler, Ch.E. Guye, W. Heisenberg, M. Knudsen, H. A. Kramers, P. Langevin, W. Pauli, M. Planck, O.W. Richardson, C.T. R. Wilson.
NB: Schrödinger's name should have been added to the list.
Secretary: J. E. Verschaffelt.

Epilogue: from "Solvay I" to "Solvay V"

Council Reports:
1. *X-ray reflection intensities*, by W. L. Bragg.
2. *Discrepancies between experiment and the electromagnetic theory of radiation*, by A. H. Compton.
3. *The new dynamics of quanta*, by L. de Broglie.
4. *Quantum mechanics*, by M. Born and W. Heisenberg.
5. *Wave mechanics*, by E. Schrödinger.

Comments

1. Mention is made in the official volume of a sixth report, allegedly presented by N. Bohr (figure 58) and entitled "*The quantum postulate and the new development of atomistiscs*". This point is surprising since we know that Bohr wasn't expected to present a report. An explanation is that it seemed appropriate to the Scientific Committee to present, as a Solvay report, Bohr's contribution to the Como conference (a distortion of the truth, reminiscent of the addition of a "report" by Kamerlingh Onnes to the minutes of the first Solvay Council).

2. Solvay V may be characterized by two contrasts: reconciliation of physicists who had been separated by the war, against a "clash" between the representatives of two generations: the *"boy's club"* represented by Pauli, Heisenberg and Dirac, *versus* the *"old guard"* represented by Lorentz, Planck, Einstein, and Schrödinger.

On the other hand, one could say that the debates were dominated by Bohr's notion of *complementarity*, a generalization of Heisenberg's *indeterminacy principle*[585].

The idea of a fundamental *indeterminacy* at microscopic level, a central subject in the conference, was based on recent experimental results: Compton's *effect*[586], the findings of Davisson and Germer which established the existence of *matter waves*, predicted by L. de Broglie (and confirmed by W. Elsasser's analysis[587]).

After having heard Bragg's arguments in favour of classical optics, the Council was informed by Compton (figure 59) of the recent evidence in support of the photon-concept.

FIGURE 58: Niels Bohr in 1927 (centre of the picture), see figure 57.

FIGURE 59: Arthur Holly Compton (sitting behind Einstein), see figure 57.

The discussion of Compton's report was an occasion for Lorentz to make a final plea in favour of the electromagnetic ether. It was based on the fact that the concept didn't contradict the predictions of relativity theory (a point reinforced by the conclusions of Einstein's gravity theory), and that it didn't prevent the existence of photons.

3. The Council members examined 3 proposals for a quantum theory: de Broglie's pilot-wave theory, Schrödinger's wave mechanics, Heisenberg-Born's matrix mechanics (an approach reformulated by Dirac).

They recognized the equivalence of the last two proposals, an observation which by itself signalled the birth of an all-encompassing "quantum mechanics".

4. Unlike the Como conference, which had taken place shortly before the Council, the Brussels meeting benefitted from Einstein's presence. Everyone was waiting for his reaction to the proposal of a revolutionary mechanics which, as Heisenberg put it[588], *"was based on a mathematical formalism which made impossible a simultaneous and perfectly precise description of the position and the velocity of an electron"*.

Yet, Einstein remained extremely discrete. He intervened just once, analysing a thought experiment which led him to the conclusion that quantum mechanics couldn't be considered *"a complete theory"*.

The Bohr–Einstein debate

A special event that people usually associate with Solvay V is the famous "Bohr-Einstein debate" about the status of quantum theory and Heisenberg's indeterminacy relations. Yet, no trace of this debate appears in the conference report[589].

It is clear that lively discussions between Bohr and Einstein took place outside the conference room – in the morning at breakfast, or in the evening at the hotel. This is corroborated by several testimonies[590], in particular by Bohr and by Ehrenfest, who witnessed Einstein's discomfort at the prospect of a definitive abandonment of causal descriptions in space and time of microscopic phenomena.

Bohr recalled[591] in 1962 the discussions he had with Einstein during the fifth Solvay meeting, in the presence and under the influence of Ehrenfest. He also specified the question which constituted the focus of their debate: *"Should one consider the description provided by quantum mechanics as exhausting all possibilities of accounting for observable phenomena, or should one, as argued by Einstein, push the analysis further so as to obtain a more complete description of these phenomena?"*.

Bohr also stressed that their disagreement took a new, more dramatic turn in 1930, at the sixth Solvay meeting. It was during this last Council attended by Einstein that Bohr succeeded in detecting, after a restless night, the flaw in Einstein's attempt to imagine a scenario[592] that would defeat the indeterminacy principle.

It is worth emphasizing Bohr's awareness of the impact of his Solvay confrontations with Einstein. Indeed, more than three decades later, the leader of the Copenhagen School underlined their influence on the evolution of his thought, recognizing that his discussions with Einstein enabled him to understand the very meaning and scope of the complementarity principle.

Einstein's unexpected encounters at Solvay V

It was on the sidelines of the fifth Solvay Council that Georges Lemaître made Einstein's acquaintance. Here are some lines of a paper[593] published in 1957 by the father of the Big Bang hypothesis:

> "I met Einstein for the first time some twenty-nine years ago. He had come to Brussels to attend the Solvay Congress of 1927. While walking in the alleys of Parc Léopold he told me about a little noticed paper that I had written the previous year on the expanding universe, and that a friend had shown to him.
>
> After a few favourable remarks of a technical nature, he concluded by saying that it seemed to him as totally abominable from the physical point of view... As I tried to extend the conversation, I was invited by Auguste Piccard, who accompanied Einstein in his walk, to enter a cab which would enable us to visit his laboratory at Brussels' University. In the cab, I talked about the velocities of nebulae and got the impression that Einstein was hardly aware of the astronomical facts
>
> At the University everything happened in German, and I was surprised to hear myself introduced as "Herr Lemaître". I admired the interferometer which had just made a balloon ascent, and signed after Einstein, the University's guest book...".

Another noteworthy event happened on the last day of the Conference, a few hours before Einstein's visit to his Belgian family.

On October 28, Einstein sent a letter to his uncle Caesar Koch, who lived in Liège with his daughter Suzanne, to tell him that he would visit him the next day. Not knowing what time the Council would end, he asked not to wait for him at the station[594]. The warning was wise, for we know that the Solvay deliberations lasted until 5:00 in the afternoon.

On October 29, Einstein was one of seven Council members, invited for lunch at the Palace with the royal couple. As the only German he was seated to the right of Queen Elizabeth. A friendly relationship immediately developed between these two exceptional representatives of southern Germany. So, we can say that the fifth Solvay Council marked the beginning of Einstein's friendship with the Queen, a relationship which lasted until his death.

9.4 Some final thoughts

Before closing this section, devoted to the three last Councils chaired by Lorentz, we should stress once more ISIP's exceptional privilege of surviving the First World War. This survival was due, as we said, to the efforts of the Institute's scientific director and to Solvay's conciliatory attitude.

The fifth Council was a milestone. After thirteen years of isolation German physics was again represented. Its impact was impressive. Lorentz was given the chance to crown his international career by welcoming German and Austrian researchers who amazed the scientific world with the power and novelty of their most recent results (Pauli's exclusion principle, Heisenberg's indeterminacy relations, matrix mechanics of Born and Heisenberg, Schrödinger's wave mechanics).

The decisive role of the Solvay debates of 1927 was recognized thirty years later by Heisenberg, who declared[595]:

"When one tells the development of quantum theory, one must particularly underline the discussions which took place in Brussels, under Lorentz's chairmanship, at the Solvay Council of 1927. By allowing representatives of different approaches to exchange on the spot, this conference contributed in an extraordinary way to the clarification of the physical foundations of quantum theory. One could even say that it constituted, in a way, its completion."

It is important to recall that at the origin of this brilliant success, and of the impact of later Councils (chaired by other eminent personalities, such as Paul Langevin, Sir Lawrence Bragg, Robert Oppenheimer, Edoardo Amaldi), there was the passionate will of a man who had the chance (if not the intuition) to take action at the very moment when physics was facing the greatest challenge in its history.

It is most remarkable that Solvay resumed his own research in 1910, that he immediately responded to Nernst's call, and that he benefitted in 1912 from Lorentz's invaluable assistance. Equally remarkable are the personal motives which prompted him to intervene in 1911 and to launch the foundation of an International Physics Institute: his unwavering confidence in a *Gravito-Materialitic* theory, and his compelling desire to confirm its predictions by means of new experiments.

However, there is a puzzling peculiarity which must be underlined. From 1912 Solvay found himself in a position to submit his ideas to a board of eminent physicists, and to benefit from their expert advice. In view of this situation, there is a question that we should ask ourselves: *What about the members of the ISC who were particularly close to the industrialist (we think in particular of Lorentz, Marie Curie and Kamerlingh Onnes)? Why didn't they try to enlighten him by informing him of their reservations about his theories?*

Epilogue: from "Solvay III" to "Solvay V"

Let us note that for this to happen it would have been necessary for Solvay to consult his professional partners. But we know that he didn't do so.

As already pointed out, the industrialist wanted to maintain a strict partition between ISIP's official activities and his personal research, thus reproducing the separation he had established between the two Physiology Institutes at Parc Léopold (one part of the building housed an Institute directed by the University, *in which Solvay had nothing to see, nor to say*, the other part housed Solvay's own Institute in which the University *had nothing to see, nor to say*).

This attitude didn't fail to amaze those around him. Let us recall Sommerfeld's words in his letter to his wife on October 31, 1911: *"Mr. Solvay told us about his discoveries with great tact, but making sure to cut short any discussion on the subject..."*.

On the other hand, we have seen that the industrialist didn't hesitate to consult some leading experimenters in which he confided. He repeatedly asked help from Marie Curie, submitting to her his personal views on the origin of radioactive phenomena. He also had regular exchanges with Kamerlingh Onnes[596] and took the trouble to visit him on several occasions... The two men had much in common. Both were brilliant organizers (Solvay in the industrial domain, Kamerlingh Onnes as an initiator of "big science"). Both had an interest in the production of cold and in the behaviour of matter at extremely low temperatures. As to the strength of their friendship, it clearly appears in this testimony, published by Kamerlingh Onnes after Solvay's death[597]:

"Like many others who devoted themselves to science, I had the privilege of being a friend of Solvay. Initially, our discussions focused on the expansion experiments carried out by Solvay to cool the air to a temperature of -95 °C. This very remarkable result, at the time, was the subject of one of my publications.

For Solvay it was always a great satisfaction to see that science, in its regular development, led to results, or at least to ideas, consistent with the conclusions he had obtained from personal reasoning. We can say that on this point there was a parallel between his work and mine. This was the case, much later, for superconductivity, a phenomenon of which he had foreseen the possibility.

But our real link was the common experience which brings the technician and the experimenter closer. That Solvay felt this was clear to me on the occasion of his visits to the cryogenic laboratory. The failures that I encountered in some experiments evoked in him the happy memory of the clearly more formidable obstacles which had arisen in his path and which he had overcome...".

It may be argued that these lines are from a talented experimenter (Nobel laureate in 1913), who doesn't tell us much of the reservations he may have had about Solvay's "very personal" concepts...

However, regarding this last point, we can rely on the testimony of a first-class theorist. These are Lorentz's words in a confidential letter[598] that he sent to Lefébure on September 29, 1922:

> "*Some may say to me: Why did you not submit to Mr. Solvay, during his lifetime, the doubts which his speculations raised in you? To this I must answer that he knew perfectly well that his views didn't agree with those of the majority of physicists, and that, more than once, I told him expressly that we were following very different paths. But we didn't have discussions on the details. Mr. Solvay didn't seem to want such discussions, and it seemed to me difficult, and even useless, to initiate them on my side. Useless, because I felt his convictions were unshakable.*
>
> *It is very remarkable – and one sees in it his greatness of soul – that despite his convictions, his "stubbornness" as he often called it, he did so much – through the creation of the International Physics Institute and in other ways – to encourage the research of physicists, and that he let them have such complete freedom...*".

We have proof that Lorentz recorded the above considerations in a document which served as a draft of his letter to Lefébure, and in which he announced his intention to publish, with Herzen's help, a note on Solvay's ideas and works.

Deeply moved by the industrialist's disappearance, ISIP's scientific director agreed to answer the questions of relatives who wondered whether it was necessary to publish Solvay's "*Essay on the Fundamental Synthesis of the Universe*" and his papers (*Mémoires*) deposited at Belgium's Royal Academy in sealed envelopes.

Having read these works, Lorentz told Lefébure that their publication would add nothing to the glory and the rights to universal recognition that Solvay had acquired through his actions and achievements in chemistry. Expressing a genuine affection for ISIP's founder, he added this touching confession: "*I can tell you that if Mr. Solvay had been my father, I would have chosen to drop his speculations in oblivion...*".

This opinion didn't contradict Lorentz's decision to publish, together with Herzen, a note "*On the Relationships between Energy and Mass, according to Ernest Solvay*" which appeared in the Proceedings of the Paris Academy of Science. To complete his laudatory remarks on the industrialist's greatness of soul, we chose to add these final lines, written by Kamerlingh Onnes in homage to his lost friend[599]:

> "*When laboratory researchers assisted him in carrying out an experiment, he derived as much satisfaction from his contacts with them, as from obtaining the desired results. These contacts each time brought him the*

Epilogue: from "Solvay III" to "Solvay V" 241

FIGURE 60: Members of the first Solvay Council on Chemistry (1922), Brussels, Solvay Physiology Institute, Parc Léopold, S.a.b.ULB. C Courtesy of the International Solvay Institutes for Physics and Chemistry. Touching detail: Solvay is sitting in the middle of the first rank. It was his last public appearance, he passed away on May 26, 1922, one month after the banquet offered to ISIC's guests.

pleasant confirmation that the pursuit of his experiments on the movement of air masses would lead him to the goal he had set for himself. After having indicated that he had "walked part of the way, once more, without reaching his goal", he declared, as a scientific researcher would have done: "I continue.".

This is how Solvay expressed his unwavering faith in the effectiveness of research carried out with method and with perseverance. He had the deep feeling that this was what had enabled him to write an important page in the history of chemistry "in the service of man[600]*".*

Solvay's understanding of the material concerns which weigh on all who engage in disinterested scientific experimentation, led him to lend his support whenever he saw the need. It is with gratitude that I think back to the important resources that he put at the disposal of Leiden's cryogenic laboratory, to help it in its investigations and in the training of technicians responsible for building measuring instruments.

But Solvay supported many other researchers. "*It would be stronger than me,* he said, *not to associate myself, by a form of encouragement, with such passionate labour*". *When he intervened, it was always in a discreet and moving manner, which surprised by its cordiality. All those who benefitted from his generosity will remember his eyes, in which shone the consciousness of having done what was right, as well as the slight smile which added to his features, witnesses of an unyielding will and of hard work, the revealing sign of a successful collaboration...*".

ANNEXES

Annex 1
List of 52 Nobel laureates who took part in one (or in several) Solvay Councils between 1911 and 1933, or who benefitted from a Solvay research subsidy

By order of award:

H.A. Lorentz (1902), P. Zeeman (1902); M. Curie (1903), S. Arrhenius (1903); Lord Rayleigh (1904); J.J. Thomson (1906); A.A. Michelson (1907); E. Rutherford (1908); J.D. van der Waals (1910); W. Wien (1911); V. Grignard (1912); H. Kamerlingh Onnes (1913); M. von Laue (1914); W.H. Bragg (1915), W.L. Bragg (1915); C.G. Barkla (1917); M. Planck (1918); J. Stark (1919); W. Nernst (1920); A. Einstein (1921), F. Soddy (1921); N. Bohr (1922), F.W. Aston (1922); K.M. Siegbahn (1924); J. Franck (1925), G. Hertz (1925); J. Perrin (1926); A.H. Compton (1927), C.T.R Wilson (1927), H. Wieland (1927); O. Richardson (1928); L. de Broglie (1929); W. Heisenberg (1932), I. Langmuir (1932); P.A.M. Dirac (1933), E. Schrödinger (1933); J. Chadwick (1935), F. Joliot-Curie (1935), I. Curie (1935); P. Debije (1936); E. Fermi (1938), R. Kuhn (1938); E. Lawrence (1939), L. Ruzicka (1939); G. de Hevesy (1943); W. Pauli (1945); P. Bridgman (1946);P. Blackett (1948); J.D. Cockcroft (1951), E.T. Walton (1951); M. Born (1954), W. Bothe (1954).

NB: Seven Councils on Physics and four Councils on Chemistry took place before the Second World War.

Annex 2
Archival sources relating to the works of Ernest Solvay

We have relied in what precedes on undisclosed or partially disclosed works.

Undisclosed sources cited in notes 16, 22, 44, 47, 136, 143, 153, 166, 346, 427, 498, 501, 508 and 604

These are typed notes, written from 1910 to 1921, in which Ernest Solvay takes stock of the state of progress of his investigations, and informs his collaborators of the work he expects from them. They are part of Solvay's daily register, a collection of private notes which are part of the Archives of the Belgian Chemical Society (Service des Archives et des Bibliothèques de l'Université Libre de Bruxelles, inventory numbers 16 and 17-22). The notes relevant to our work are those of the period 1910-1914.

Partially disclosed sources

i) *Notes sur les travaux poursuivis par Ernest Solvay de 1857 à 1914,* Bruxelles, Imprimerie G. Bothy, 1920. About fifty copies of this anonymous document have been privately printed. The author (obviously Emile Tassel) was careful to indicate that the work was not intended for publication. It is nonetheless cited by Louis D'Or and Anne-Marie Wirtz-Cordier in *Ernest Solvay,* Académie royal de Belgique, Mémoires de la Classe des Sciences, 2e série, T. XLIV, Fascicule 2, 1981, 96, by Isabelle Stengers in *Ernest Solvay et son temps,* Eds. A. Despy-Meyer et D. Devriese, Archives de l'Université Libre de Bruxelles, 1997, 165. and by Kenneth Bertrams, Nicolas Coupain and Ernst Homburg in *Solvay: History of a Multinational Family Firm,* Cambridge, 2013, 580. The above source, of which a copy can be found in H. A. Lorentz's Collection of the Teyler Museum in Haarlem, is cited in notes 20, 30, 48 and 53.

ii) E. Solvay, *Gravitique. De la gravité astronomique de la matière et de ses rapports d'équivalence avec l'énergie.* This work, completed in 1887 with the help of Tassel, was printed in 1895 and recorded in the archives. It is part of a collection entitled *Notes, lettres et discours d'Ernest Solvay,* vol. 1, Bruxelles, Lamertin, 1929.

A few copies of Solvay's work are kept at the head office of the Solvay Group.

The above source is cited in notes 45, 58 and 252.

iii) E. Solvay, *Sur l'établissement des principes fondamentaux de la gravito-matérialitique.* This study of 109 pages was printed by G. Bothy in October 1911.

It comprises four parts: I. *Gravitique* (continuation of the previous work); II. *Self- gravitique;* III. *Gravito-Matérialitique*; IV. *Matérialitique.*

The work was presented to the members of the first Council on Physics. A copy of Solvay's study, cited (or referred to) in notes 21 and 193, can be viewed at the Library of the Université Libre de Bruxelles.

Are available at Teyler's Foundation in Haarlem:

1. E. Solvay, *Sur l'établissement des principes fondamentaux de la gravito-matérialitique,* Bruxelles, 1911.
2. E. Solvay, *Gravitique. De la gravité astronomique de la matière et de ses rapports d'équivalence avec l'énergie.* Bruxelles, 1895.
3. Anonymous: *Notes sur les travaux poursuivis par Ernest Solvay de 1857 à 1914* Bruxelles, 1920.

Annex 3
Solvay's "Gravito-Materialitic" program

Preliminary remarks

1. Solvay was convinced, as we have underlined, that physiology and sociology (at the time a very recent discipline) were governed by the laws of physics and chemistry.

This guiding idea – basis of his research program – had been defended before him by several authors. Condorcet, an eminent representative of the Enlightenment, already intended to apply the exact sciences to the new ones[601]: sciences "the object of which is man himself". The idea was taken up by the biologist and philosopher Ernst Haeckel, whose portrait appears in the "*Ernest Solvay's Bedside Table*[602]".

Here is what Haeckel wrote in *Die Welträthsel*, a work of 1899 translated into French under the title *Les énigmes de l'Univers*, chapter XX:

> "*The great abstract law of mechanical causality...now governs the entire universe, as it governs the human brain*".

2. When he speaks of the '*positive*" Solvay is referring to a generalized physics which governs the Universe, and which, according to him, should apply to all areas of human activity. It is in this sense that he considers the scientific field to be unlimited.

3. Solvay uses in his writings terms which belong only to him: a habit that makes his reasoning difficult to follow.

Elements of comparison

We can try to situate Solvay's "*Gravito-Materialitic*" approach with respect to the dominant currents of the time. Let us first underline the leading role attributed to energy. This characteristic places his project in the wake of a well-known current: *energetics*. Among the defenders of this doctrine, we find first rank masters, such as Georg Helm, Wilhelm Ostwald and Ernst Mach.

For a whole school of scientists, *energy* – the conservation of which was established as a principle when Solvay was twenty – was much more than an abstract concept. It had an *objective* existence, as fundamental as that of matter.

Energetists adopted Helm's principle, according to which "all phenomena can be reduced to energy transformations".

When Solvay adhered to this principle, and when he wished, with Helm[603], for the emergence of an *objective* physics, free from metaphysical elements such as matter and movement, he didn't follow Ostwald and Mach in their unconditional rejection of the "atomic hypothesis". However, his views on the atom were far removed from those of the pioneers of modern atomism: he admitted the existence of *ether* atoms, invariable and cubic in shape[604] (only the molecule was in his eyes of variable volume and shape).

These are Solvay's words[605] in a letter of 1907, written in reaction to the death of Marcellin Berthelot, a well-known opponent of atomism (the letter was sent to the latter's son):

"*I was with him in the resistance he showed to the modern atomic movement, which had its raison d'être, but which will not remain, in my opinion, as the last expression of reality.*"

Yet, several elements indicate that Solvay's ideas about atoms evolved over time (see his invitation letter to the Council in which he repeated this declaration from Nernst: "*A big step in the way of the development of atomistic would already be made if we could clearly establish which of our molecular and kinetic interpretations are in agreement with the results of observation, and which, on the contrary, will have to undergo an integral transformation*").

In a note of September 15, 1910 (Archives of the Belgian Chemical Society, undisclosed source), Solvay bases his ideas about the structure of matter on geometric considerations relating to the *ether*. According to Hertz and Lord Kelvin[606], the nature of this hypothetical medium should enable us to understand the essential properties of matter, *i.e.*, its gravity and its inertia.

In France, Poincaré was one of those who wondered about the nature of *ether*[607]:

"*What is ether? How are its molecules arranged, do they attract or repel each other? We do not know, but we know that this medium transmits both optical and electrical disturbances.*"

In the United Kingdom, the physicist Karl Pearson had very precise views[608] on the subject:

"*The ether atom must be seen as the primary element of the Universe, the chemical atoms being formed from these fundamental elements, united in rings...*".

So, we can say that Solvay was in good company when he indulged in speculations on the geometric structure of the ether and on the role of its atoms.

What is less known is that Solvay's ideas about the ether aroused Planck's interest (see his comments in section 2.7 on Solvay's study: *Sur l'établissement des principes fondamentaux de la gravito-matérialitique*).

Indeed, the discoverer of energy quanta didn't hesitate to express his admiration for *Solvay's absolutely independent and original ways of seeing*, even if he criticized the fact that the industrialist's views were exclusively based on gravity.

Poincaré's (alleged) influence on Solvay

It is known that Poincaré used to wonder about the validity of generally accepted principles in physics (see his books published by Flammarion). His words seem to have found an echo in Solvay, who just like Poincaré was profoundly disturbed by the extraordinary persistence of Brownian motion, and by the "seemingly inexhaustible amount of heat" generated by radium. The industrialist seems to have been struck by Poincaré's disconcerting claim (*La valeur de la science*, p. 198):

> "*Will these principles, on which we have built everything, collapse in their turn? For some time now we may wonder...*".

We may even assume that Solvay felt reinforced in his rejection of the second law of thermodynamics by Poincaré, who declared that Carnot's principle appeared to him as "*a concession to the infirmity of our senses*".

In conclusion, we can say that Solvay's approach answered a legitimate questioning at the time, and that his concerns were those of some leading physicists. The main factor which differentiates him from professional scientists is his belief that he can build by mere reflection, and without any real training in physics, a valid and all-encompassing theory of the "active Universe". This illusory idea seems to have acquired in him the strength of a conviction, of a stubbornness which pushed him to embark on more and more exotic paths, at the risk of getting lost... But should we condemn an obstinacy which enabled this visionary to contribute, through lasting international foundations, to the progress of physics and chemistry?

Annex 4
The Black-Body Problem

What is meant by "The problem of black-body radiation"?

In his 1913 address of homage to Planck, published in *Naturwissenschaften*[609], Einstein reminds us that each body emits thermal radiation, so that a cavity inside a material enclosure is constantly interspersed by this radiation. He then notes that Kirchhoff showed around 1860, with the help of thermodynamic considerations, that the thermal radiation is uniform in all directions and that its properties can only depend on the enclosure's temperature T. Thus, for a given frequency v, the radiation energy per unit volume is determined by temperature T and by the value of the frequency. But the energy does not depend on the physical or chemical nature of the enclosing wall.

This means that the energy distribution is a universal function of two variables: frequency v and temperature T.

Einstein continues by reminding us that the determination of the function f(v,T) became the main task of the theorists who worked on the problem. He first mentions the progress realized by Boltzmann, who in 1884 found the variation with temperature of the total energy distribution (*i.e.*, integrated over all frequencies). He then follows with Wien's law of 1893, which established that the sought function f only depended on one variable: the ratio v/T of frequency and temperature.

Einstein finally points out that Planck was the first physicist who managed to find the function f(v/T), and who presented a ground-breaking theoretical way to derive it. He concludes his summary by describing the steps which enabled Planck to express the function f(v/T) in terms of two universal constants: the already known Boltzmann constant and a new one, Planck's constant. The resulting expression f(v/T) became famous as "*Planck's radiation law*".

Annex 5
Planck's "missed" Nobel Prize

We have seen that the news had spread that Planck would get the 1908 Nobel Prize in Physics. Here is an excerpt from the interview he gave on this occasion, and which was published in *Leipziger Neueste Nachrichten*[610] on November 23, 1908.

> "*I presume that I owe this honour principally to my works in the area of heat radiation. For some time I have been occupied with establishing the absolute weight of the atom, and I have now succeeded in arriving at a positive result*[611].
>
> *It was earlier assumed that the absolute weight of the atom couldn't be established. We are now out of the area of speculation and have reached firm ground... What is interesting here is that Rutherford, who is primarily concerned with research on radioactive bodies and substances, has arrived, through means quite different from mine, at approximately the same value*[612]...".

Annex 6
The second Moroccan crisis and the Caillaux affair[613]

Morocco aroused since the beginning of the 20th century, the greediness of several European powers, notably of France and the German Empire (the latter feeling the need to catch up in colonization).

In 1904, France and the United Kingdom concluded an agreement against Germany.

France would leave Britain's hands free in Egypt; in return, it would install a protectorate in Morocco. In March 1905, Kaiser Wilhelm II landed in Tangier in order to remind of his claim on Morocco. He had an encounter with Sultan Moulay-Abd-al Aziz. This was the first Moroccan crisis (the Tanger crisis), which provoked tensions between the European powers, and led in 1906 to the Algeciras International Conference.

In March 1911, Sultan Moulay ab-Hafid, under threat of revolt, requested military assistance from France. In the month of May, French troops occupied Rabat, Fez and Meknes. Germany's reaction was fierce, as it considered this occupation as a violation of the 1906 Algeciras agreements. From July 1, 1911, it sent gunboats taking turns in the bay of Agadir. This was the subject of a serious crisis between France and Germany. The two countries were at the edge of an armed conflict.

In France, the government had just changed. Joseph Caillaux, President of the Government's Council since June 27, 1911, wanted to avoid war. He wished to negotiate with Germany on France's freedom of handling in Morocco in exchange for the cession to Germany of French territories in Central Africa. Yielding to the pressure from Russia and from the United Kingdom, an agreement between France and Germany was reached on November 4, 1911 (French-British negotiations regarding the treaty of Fez). Germany renounced its aspirations in Morocco in exchange for France's abandonment of territories in Gabon, Middle Congo and Ubangi. But the agreement provoked concern to King Albert of Belgium, for it meant the establishment of German settlements in front of the Belgian territories in

Congo (it was no secret that Germany regarded Belgium as a country too small to govern this huge territory...).

The issue flared up again in France, as President Caillaux bypassed his Minister of Foreign Affairs, Justin de Selves, and addressed directly Alfred von Kinderlen-Waechter, the German Minister of Foreign Affairs. Serious Problem: De Selves learned that secret negotiations were going on between France and Germany (the French Department had succeeded in intercepting and deciphering Germany's encrypted messages[614]). Feeling humiliated, he provoked with the help of Clemenceau, Caillaux's old enemy, an incident during the ratification in the Senate of the French-German agreement (January 9, 1912).

Final development: Raymond Poincaré succeeded Caillaux and got the treaty of November 1911 accepted.

Annex 7
Royal patronage

On September 2, 1912, Solvay sent the following message[615] to general Jungbluth:

> My dear General,
>
> I allow myself to send you a copy of the statutes, not yet final, of the International Solvay Institute for Physics, which I founded in Brussels in agreement with Professor H. A. Lorentz from Leiden, following the Physics Council which was held here at the end of last October, and about which I had, at that time, an interview with you... You will see in these statutes that it is envisaged that a delegate to the Administrative Commission will be appointment by the King. It is Mr. Lorentz who introduced the idea, and I will have to ask you, in case you kindly grant me an audience, if you believe that His Majesty would agree to follow up on the proposal, something for which all members of the Institute would feel highly honoured, your servant in particular...

Following the constitution of the International Scientific Committee, Solvay found himself in a position to provide further clarifications. On September 20, 1912, he addressed another message[616] to Jungbluth:

> My dear General,
>
> In order to be able to use it, if necessary, I believe I should inform you on the situation of the International Physics Institute from the point of view of its two governing bodies. The international Scientific Committee has decided to include Mr. Robert Goldschmidt as a member (his adhesion to the proposal must still be requested). As to the Administrative Commission, I intend to suggest to the University to choose for its delegate Mr. Verschaffelt, professor of physics and member of the Academy of Sciences. My personal delegate would be my long-term collaborator

Mr. Tassel, honorary professor at the University. These proposals are still confidential, but there is general agreement on the names. In case His Majesty would accept the situation that the Scientific Committee has in view[617], I wonder if, among the delegates that He could think of, I could suggest that you put on the list Professor Héger, President of the Academy of Medicine and director of the Physiology Institute? I am sure that this choice would be unanimously appreciated...

Jungbluth answered[618] on October 1, 1912:

"*The King, who has taken note of all the documents concerning the International Physics Institute, instructs me to congratulate you very warmly on your generous and fruitful initiative. He is happy to respond to the wish of the International Scientific Committee by accepting the mission of appointing a member of the Institute's Administrative Commission. His Majesty has fixed his choice on Monsieur Héger, your former and devoted collaborator, whose knowledge and character command everyone's esteem...*".

Annex 8
Essential points in the Rutherford-Thomson confrontation

J. J. Thomson:

> "*From time to time we see that the direction in which the alpha particles move changes abruptly, as if their trajectory had been deviated over a finite angle by a single encounter with a molecule. The magnitude of the deviation proves that, if it results from the electric forces exerted on the alpha particle by a charged body with which it collides, both the charge and the mass of that body must be large in comparison to the corresponding amounts for the alpha particle. This led Professor Rutherford to consider the entire positive charge and practically the whole mass of the atom as concentrated in an excessively small volume at its centre, that is to say that the radius of the region in which the mass is assumed to be concentrated is excessively small compared to the conventional radius of the atom. The repulsion of this great charge on a nearby passing alpha particle is considered to be the cause of the irregularities which appear at intervals in the trajectories of these particles.*
>
> *If these irregularities in the trajectory of the alpha particles are due to repulsions of a charge equal to the total charge of all atomic corpuscles (electrons), concentrated at a point, the electric field from this charge must produce in the trajectory of a negatively charged particle a much greater number of breaks than in that of a positively charged particle, since the mass of the negative particle (the electron) is by far the smaller of the two... But the photographs that C. T. R. Wilson took of the trajectories of the negative particles emitted by the air molecules when they are exposed to Röntgen rays show no sign of abrupt changes in direction of the corpuscles of a size approaching that indicated in the table* (Thomson presents a table which shows the sizes of angles that should be observed in this case).

This seems to me to indicate[619] *that the great changes, which from time to time occur abruptly in the direction of movement of alpha particles, are not produced by forces due to electric charges*[620]*... In my opinion, they must rather be attributed to special forces that come into play when two alpha particles approach each other at a distance less than a certain limit. I think that in reality, when two alpha particles collide inside an atom, the forces acting between them are not only those which would be exerted between the particles under the ordinary electrostatic laws. Besides these forces, there are others*[621] *which make themselves felt, and it is they which produce the characteristic irregularities in the trajectories of the alpha particles."*

Rutherford:

"Sir J. J. Thomson drew our attention to the important fact that the traces of beta particles fixed by photographic means by C. T. R. Wilson do not present the many large deviations that we would expect from the theory of the atomic nucleus that I put forward some time ago. I would like to insist on the precision of the experimental evidence on which this theory is based... It has been assumed that the large deviations of alpha particles (of great velocity) are due to the passage of these particles through the intense electric field in the immediate vicinity of the central atomic nucleus. A careful comparison of theory and experiment has been made by Geiger and Marsden, and the deductions drawn from the theory have been found to be in perfect agreement with the experimental results. We could deduce that the charge of the nucleus, and therefore the number of electrons in the atom, is approximately equal to half the atomic weight of all the elements examined. The force acting between the alpha particle and the nucleus is supposed to follow the ordinary inverse square law, and it is found that any other law of force law is incompatible with experimental facts.

One would of course expect that the high-speed (beta) electrons, having about the same energy as the alpha particle, also undergo large abrupt deviations as they pass very close to the nucleus. I will however recall that, while the energy of the beta particle considered by Sir J. J. Thomson is much smaller than that of the alpha particle, this beta particle is supposed to pass through the intense field produced by the central nucleus. Now I think that is it doubtful that the beta particle at low speed penetrates into the region under consideration, where the alpha particle undergoes large deflections; it seems to me rather that it will remain in the region of the atom where the field produced by the nucleus is very weakened, and where it will undergo small deviations as a result of the shocks against the electrons associated with the atom. The question whether a high-speed beta particle, of energy comparable to that of the alpha particle, undergoes large abrupt deviations is of very great theoretical importance and is being examined experimentally... Unless it is assumed that atoms have a charged nucleus of small dimension, it is impossible to explain experimental facts of alpha particle scattering without admitting that new and unsuspected forces of attraction and repulsion, of a very intense nature, act between atoms when they are very close. If we consider that the theory of the nucleus leads to the same number of electrons in an atom as the scattering of X-rays, it seems to me simpler to suppose that such a charged nucleus has a real existence and that it is a fundamental constant of all atoms."

Bibliography

I. Works classified by authors' names (the list only includes works cited several times).

G. Bacciagaluppi and A. Valentini, *Quantum Theory at the Crossroads: Reconsidering the 1927 Solvay Conference*, Cambridge University Press, 2009 (564, 566, 584, 589).

H.-G. Bartel and R. P. Huebener, *Walther Nernst: Pioneer of Physics and Chemistry*, World Scientific, 2017 (483, 484).

F. Birck, *Des ingénieurs pour la Lorraine*, Edition Serpenoise, 1998 (179, 412).

N. Bohr, *The Solvay Meetings and the Development of Quantum Physics*, dans *La Théorie Quantique des Champs* (Douzième Conseil de Physique), R. Stoops (Ed.), New York, Interscience Publishers, 1962 (240, 591, 592).

A. Boutaric, *La lumière et les radiations invisibles*, Paris, Flammarion, 1925 (49, 51).

R. Brion and J. L. Moreau, *Franz Philippson. Aux origines de la Banque Degroof*, Bruxelles, D. Devillez, 2016 (119, 484, 485).

R. Clark, *Albert Einstein. The Life and Times*, London, 1973 (272, 273, 274, 312, 317).

P. Coffey, *Cathedrals of Science. The Personalities and Rivalries that Made Modern Chemistry*, Oxford University Press, 2008 (564, 574, 575, 576, 577).

N. Coupain, *Ernest Solvay's scientific networks. From personal research to academic patronage*, in *The Early Solvay Councils and the Advent of the Quantum Era*, F. Lambert, F. Berends and M. Eckert (Eds.) Eur. Phys. J. ST, vol. 224, 10, 2015 (46, 56).

E. Crawford, *The Beginnings of the Nobel Institution. The Science Prizes, 1901-1915*, Cambridge University Press, 1987 (103, 106, 110, 112, 114, 245, 610).

M. de Broglie, *Registre contenant des pièces manuscrites concernant les premiers Congrès Solvay (2 volumes) offert à l'Académie des Sciences*, Archives de l'Institut de France, 1951 (203, 218).

J. Dhombres and J. B. Robert, *Fourier créateur de la physique mathématique*, Paris, Belin, 1998 (419, 601).

L. D'Or and A.-M. Wirtz-Cordier, *Ernest Solvay*, Académie Royale de Belgique, Mémoires de la Classe des Sciences, T. XLIV, Fascicule 2, 1981 (27, 28, 29, 31, 33, 42).

A. Einstein, *L'état actuel du problème des chaleurs spécifiques* in *La théorie du rayonnement et les quanta. Rapports et discussions de la réunion qui s'est tenue à Bruxelles du 30 octobre au 3 novembre 1911 sous les auspices de M. E. Solvay*, P. Langevin and M. de Broglie (Eds.), Paris, Gauthier-Villars, 1912 (221, 223, 224, 256).

A. Eve, *Rutherford*, Cambridge University Press, 1939 (116, 231, 247).

F. Giroud, *Une femme honorable. Marie Curie, une vie,* Paris, Fayard, 1981 (244, 251).

S. Grundmann, *The Einstein Dossiers,* Berlin, Springer, 2004 (78, 309, 310, 312).

E. Gubin, *Dans la Presse. Marie Curie et les premiers Conseils Solvay* in *Les premiers Conseils Solvay et les débuts de la physique moderne,* P. Marage and G. Wallenborn (Eds.), Editions de l'Université Libre de Bruxelles, 1995 (243, 246).

J. L. Heilbron, *The First Solvay Council « A sort of Private Conference » in The Theory of the Quantum World*, Proceedings of the 25th Solvay Conference in Physics, D. Gross, M. Henneaux and A. Sevrin (Eds.), World Scientific, 2013 (11, 15).

J. J. Heirwegh and M. Peeters, *Le tableau de chevet d'Ernest Solvay* in *Ernest Solvay et son temps*, A. Despy-Meyer and D. Devriese (Eds.), Bruxelles, Archives de l'Université Libre de Bruxelles, 1997 (41, 602).

M. Jammer, *The Conceptual Development of Quantum Mechanics,* The History of Modern Physics 1800-1950, vol. 12, AIP, Tomash, 1989 (14, 102).

H. Kamerlingh Onnes, *Ter Herdenking Ernest Solvay,* De Ingenieur 37, 1922 (60, 597, 599).

M. Klein, *Einstein, Specific Heats and the Early Quantum Theory,* Science 148, 1965 (11, 82).

M. Klein, *Paul Ehrenfest. The Making of a Theoretical Physicist,* Amsterdam, North-Holland, 1970 (71, 293).

M. Klein, *Thermodynamics in Planck's Work*, History of Physics, S. R. Weart and M. Phillips, Readings from Physics Today, New York, AIP, 1985 (159, 208).

D. Kormos Barkan, *The Witches' Sabbath: The First International Solvay Congress* in *Einstein in Context*, M. Beller, R. S. Cohen and J. Renn (Eds.), Cambridge University Press, 1993 (11, 91).

D. Kormos Barkan, *Walther Nernst and the Transition to Modern Physical Science*, Cambridge, 1999 (111, 135, 242).

A. J. Kox, *Einstein, Specific Heats and Residual Rays: The History of a Retracted Paper* in *No Truth Except in the Details*, A. J. Kox and D. Siegel (Eds.), Boston Studies in the Philosophy of Science, Kluwer, 1995 (147, 149, 306).

T. Kuhn, *Black-Body Theory and the Quantum Discontinuity, 1894-1912*, University of Chicago Press, 1978 (63, 70, 71, 90, 148).

A. Langevin, *Paul Langevin et les congrès de physique Solvay*, La Pensée, Revue du Rationalisme moderne, Paris, 1966 (10, 460, 551, 554).

F. Levie, *L'homme qui voulait classer le monde*, Bruxelles, Impressions Nouvelles, 2006 (114, 120, 173).

W. **Nernst**, *Sur les chaleurs spécifiques aux basses températures et le développement de la thermodynamique*, J. Phys. 4, t. IX, 1910 (101, 123).

W. **Ostwald**, *Lebenslinien*, Berlin, 1927 (35, 115).

W. **Ostwald**, *Denkschrift über die Gründung eines internationalen Institutes für Chemie*, Memorandum, 1912 (180, 320).

A. **Pais**, *Subtle is the Lord...*, Oxford University Press, 1982 (98, 229)

A. **Pais**, *Niels Bohr's Times*, Oxford, Clarendon Press, 1991 (365, 399, 543).

J. **Pelseneer**, *Historique des Instituts internationaux de Physique et de Chimie Solvay, depuis leur fondation jusqu'à la deuxième guerre mondiale,* unpublished manuscript, available on microfilm. Archive for History of Quantum Physics, Library of the American Philosophical Society (125, 496, 511). A copy of the manuscript is kept at the Université Libre de Bruxelles, S.a.b.ULB.

H. **Poincaré**, *La valeur de la science,* Paris, Flammarion, 1905 (43, 44, 607).

J. S. **Rowlinson**, *Sir James Dewar, 1842-1923. A Ruthless Chemist*, London, Routledge, 2016 (49, 56, 60).

L. **Rosenfeld**, *La première phase de l'évolution de la théorie des quanta*, Osiris 2, 1936 (11, 63).

A. **Schirrmacher**, *Who made quantum theory popular with physicists and beyond?* in *The Early Solvay Councils and the Advent of the Quantum Era*, F. Lambert, F. Berends and M. Eckert (Eds), Eur. Phys. J. ST, vol. 224, 10, 2015 (12, 130).

E. **Solvay**, *Sur l'établissement des principes fondamentaux de la Gravito-matérialitique*, Bruxelles, G. Bothy, 1911 (21, 193).

E. **Solvay**, *Lettres à Marie Curie*, BNF, Gallica, *Pierre et Marie Curie Papiers II*, docs. 266-269 et 275-279 (353, 354, 390).

E. **Solvay**, *Energie radioactive de transformation*, Note sent to Marie Curie on April 28, 1914, BNF, Gallica, *Pierre et Marie Curie, Papiers II*, doc. 273-274 (353, 393, 394).

E. **Solvay**, Gravitique. *De la gravité astronomique de la matière et de ses rapports avec l'énergie.* Notes, lettres et discours d'Ernest Solvay, vol. 1 Gravitique et Physiologie, Bruxelles, Lamertin, 1929 (45, 58, 252). A first version of this work, written by E. Tassel, was printed in 1887.

F. **Stockmans**, *Goldschmidt (Robert-Benedict)*, Biographie Nationale, t. XLII, Académie Royale de Belgique (54, 118, 122, 388, 487, 524).

E. **Tassel**, *Notes sur les travaux poursuivis par Ernest Solvay de 1857 à 1914*, Bruxelles, Imprimerie G. Bothy, 1920 (20, 30, 48, 53, 58).

S. L. **Wolff**, *Physiker im « Krieg der Geister »*, Historical Studies in the Physical and Biological Sciences 33, 2003 (376, 462).

P. **Zeeman and A. D. Fokker**, *H. A. Lorentz Collected Papers* Den Haag, 1934-1939, vol. 1-9 (68, 69, 72, 304, 383, 400, 401).

II. Collective works, classified by authors names.

M. Beller, R. S. Cohen and J. Renn (Eds.), *Einstein in Context*, Cambridge University Press, 1993 (11).

A. Despy-Meyer and D. Devriese (Eds.), *Ernest Solvay et son temps*, Bruxelles, Archives de l'Université Libre de Bruxelles, 1997 (41, 418).

A. Eucken (Ed.), *Die Theorie der Strahlung und der Quanten. Mit einem Anhänge über die Entwicklung der Quantentheorie vom Herbst 1911 bis zum Sommer 1913*, Halle, Wilhelm Knapp, 1914 (398).

D. Gross, M. Henneaux and A. Sevrin (Eds.), *The Quantum Structure of Space and Time*, Proceedings of the 23d Solvay Conference in Physics, World Scientific, 2007 (11).

D. Gross, M. Henneaux and A. Sevrin (Eds.), *The Theory of the Quantum World*, Proceedings of the 25th Solvay Conference in Physics, World Scientific, 2013 (11).

A. J. Kox and D. M. Siegel (Eds.), *No Truth Except in the Details*, Boston Studies in the Philosophy of Science, Kluwer, 1995 (147).

A. J. Kox (Ed.), *The Scientific Correspondence of H. A. Lorentz 1*, New York, 2008 (291, 456, 459, 463, 467).

F. Lambert, F. Berends and M. Eckert, *The Early Solvay Councils and the Advent of the Quantum Era*, Eur. Phys. J. ST, vol. 224, 10, 2015 (12, 46, 197).

P. Langevin and M. de Broglie (Eds.), *La théorie du rayonnement et les quanta, Rapports et discussions de la réunion qui s'est tenue à Bruxelles du 30 octobre au 3 novembre 1911 sous les auspices de M. E. Solvay*, Paris, Gauthier-Villars 1912 (132, 219, 221, 223, 232, 256, 541).

P. Marage and G. Wallenborn (Eds.), *Les Conseils Solvay et les débuts de la physique moderne*, Editions de l'Université Libre de Bruxelles, 1995 (11, 243).

P. Marage and G. Wallenborn (Eds.), *The Solvay Councils and the Birth of Modern Physics*, Basel, Birkhaüser, 1999 (6).

R. Stoops (Ed.), *La Théorie des Champs (Douzième Conseil de Physique)*, New York, Interscience, 1962 (240).

III. Reports presented and discussed at the Solvay Councils, organized by the International Physics Institute during the period 1913-1927.

Solvay II:

La structure de la matière. Rapports et discussions du Conseil de physique tenu à Bruxelles du 27 au 31 octobre 1913 sous les auspices de l'Institut international de physique Solvay, Paris, Gauthier-Villars, 1921 (238, 378, 384).

Solvay III:

Atomes et électrons. Rapports et discussions du troisième Conseil de physique tenu à Bruxelles du 1er au 6 avril 1921 sous les auspices de l'Institut international de physique Solvay, Paris, Gauthier-Villars, 1923 (532).

Bibliography

Solvay IV:

Conductibilité électrique des métaux et problèmes connexes. Rapports et discussions du quatrième Conseil de physique tenu à Bruxelles du 24 au 28 avril 1924 sous les auspices de l'Institut international de physique Solvay, Paris, Gauthier-Villars, 1927 (546).

Solvay V:

Electrons et Photons. Rapports et discussions du cinquième Conseil de physique tenu à Bruxelles du 24 au 29 octobre 1927 sous les auspices de l'Institut international de physique Solvay, Paris, Gauthier-Villars, 1928 (584).

Acknowledgments

We wish to thank the *International Institutes for Physics and Chemistry, founded by Ernest Solvay*, and in particular the Institutes' President, Mr. Jean-Marie Solvay, and their Director, Professor Marc Henneaux, for their constant support.

We are grateful to Mrs. Marina Solvay for providing us with family documents that allowed us to deepen Ernest Solvay's thoughts and the significance of his scientific foundations.

We address special thanks to Professor John L. Heilbron, for having encouraged us to join efforts, and to collect the archival sources (Dutch and Belgian) which illustrate the role and merits of our two heroes: H. A. Lorentz and E. Solvay.

We are grateful to Professor Diana Kormos Buchwald, to Mrs. Danielle Fauque, and to Mr. Nicolas Coupain, for attracting our attention to illuminating documents, such as Walther Nernst's correspondence with Arthur Schuster, Ernest Solvay's *daily register* and his correspondence with Marie Curie.

We wish to thank the following relatives of participants in the first Solvay Council:
- Professor Mary Fowler for offering us a copy of Arthur Eves's biography of Ernest Rutherford.
- Mrs. Nathalie Ferrard and Mr. Jean Ferrard for providing us with letters from Albert Einstein to his Belgian family.
- Mrs. Monika Baier for sending us letters from Arnold Sommerfeld to his wife.
- Mr. Pierre Verhas for providing us a with a picture of Georges Hostelet.

We are indebted to the following Directors of archival centers for their welcome and their precious support:
- Mrs. Catherine Kounelis from the *Ecole supérieure de physique et de chimie industrielles de la Ville de Paris* (ESPCI), who provided us with essential sources.
- Mrs. Anne Chevallier from the *Archives du Collège de France*.
- Former Director Didier Devriese, as well as Mrs. Carole Masson, Mrs. Françoise Delloye and Mrs. Pascale Delbarre from the *Service des Bibliothèques et Archives de l'Université Libre de Bruxelles* (S.a.b.ULB).

- Mrs. Godelieve Bolten, from the *Noord-Hollands Archief* in Haarlem.
- Mrs. Dalila Wallé from *Rijksmuseum Boerhaave* in Leiden.
- Mr. Marijn van Hoorn from Teylers Museum in Haarlem.
- Dr. Vera Enke, from the *Archiv der Berlin-Brandenburgischen Akademie der Wissenschaften*.
- Mrs. Stéphanie Manfroid from the *Mundaneum* in Mons.
- Mr. Dirk van Delft, former Director of *Rijksmuseum Boerhaave*.
- Dr. Roni Grosz from the *Albert Einstein Archives* at the *Hebraic University of Jerusalem*.
- Mr. Hippolyte Bailli, from the Archives of the *Belgian Municipality of Ixelles*.

We also wish to thank the many people who helped us to gather important pieces of information. In particular, the following ladies and gentlemen: Brigitte Van Tiggelen, Françoise Levie, Yoanna Alexiou, Martha Cecilia Bustamante, Chantal Forget, Michael Eckert, Dominique Lambert and Erik Langlinay.

Last but not least we would like to express our gratitude to those who assisted us in the production and illustration of the original manuscript of our book in French: Mrs. Dominique Bogaerts and Mrs. Isabelle Van Geet, ISIPC office managers, and our colleagues Jacques Bijtebier and Bruno Van Bogaert.

One of us (FL) has a special thought for the late Mrs. Nadine Galland, who accompanied him during many years in his archival research. He also thanks Mr. Clovis Neves for helping him to overcome the many technical difficulties specific to an increasingly digital world.

Notes

1. Originally: "Institut international de physique Solvay" and 'Institut international de chimie Solvay".

2. This Congress, which brought together 750 participants from 24 countries, was to cover all areas of physics. Many of the participants were teachers (in 1900 there were between 1200 and 1500 physicists attached to universities or polytechnics, of which a quarter in the United States: Helge Kragh, *Quantum Generations* Princeton University Press, 1999, 13.

3. Founded in 1922 the International Union of Pure an Applied Physics (IUPAP) only started its regular activity after the Second World War.

4. The study of elementary particles, which should have taken place in October 1939 (Solvay VIII) was addressed in 1948 during the first Council after WWI. Cosmology was the subject in 1958 of Solvay XI, attended by leading specialist: Georges Lemaître, Fred Hoyle, John Archibald Wheeler and Robert Oppenheimer.

5. See Annex 1, with the list of 52 Nobel laureates who took part in one, or several Solvay Councils between 1911 and 1933, or who were the beneficiaries of a Solvay subsidy.

6. Elisabeth Crawford, The Solvay Councils and the Nobel Institution, see Pierre Marage and G. Wallenborn (Eds.): *The Solvay Councils and the Birth of Modern Physics*, Basel, Birkhaüser, 1999, 43-54.

7. The picture was probably taken on October 31, 1911, after the discussion of Planck's report: we notice the "black-body radiation law", written on the blackboard, in the back of the room. The photo was clearly retouched so as to show Solvay, who wasn't present at the time (we know his decision not to attend the Council's sessions).

8. These were Solvay's words when he announced the holding of the 1911 Council to king Albert (FSI, S.a.b.ULB, doc. 1737).

9. Regarding the importance of hotel Métropole, see the article by Stéphane Foucart in "Le Monde" of July 31, 2015, "Ces hôtels qui ont changé le monde".

10. Maurice de Broglie, *Les premiers Congrès de physique Solvay et l'orientation de la physique depuis 1911*, Paris, Albin Michel, 1951; André Langevin, Paul Langevin et les Congrès de physique Solvay, *La Pensée, Revue du rationalisme moderne*, 1966. Jagdish Mehra, *The Solvay Conferences on Physics. Aspects of the Development of Physics since 1911*, Dordrecht, D. Reidel, 1975.

11. Léon Rosenfeld, La première phase de l'évolution de la théorie des quanta, *Osiris*, 2, 1936, 149-196; Martin Klein, Einstein Specific Heats and the Early Quantum Theory, *Science* 148, 1965, 173-180; Diana Kormos Barkan, The Witches' Sabbath: The First International Solvay Congress, in M. Beller, R. S. Cohen, and J. Renn (Eds.), *Einstein in Context*, Cambridge University Press, 1993, 59-82. Also: P. Marage and G. Wallenborn (Eds.), *Les Conseils Solvay et les débuts de la physique moderne*, Editions Université Libre de Bruxelles, and Peter Galison, Solvay redivivus, in D. Gross, M. Henneaux, A. Sevrin Eds), *The Quantum Structure of Space and Time, Proceedings of the 23d Solvay Conference in Physics*, World Scientific, 2007, 1-18; J. L. Heilbron, The First Solvay Council, a sort of Private Conference, in D. Gross, M. Henneaux, A. Sevrin (Eds.), *The Theory of the Quantum World, Proceedings of the 25th Solvay Conference in Physics*, World Scientific, 2013, 1-16.

12. Frederick Lindemann, Ein eigenartiger Congress, *Berliner Tageblatt*, February 12, 1912. See Arne Schirrmacher, Who Made Quantum Theory Popular with Physicists and Beyond?, in F. Lambert, F. Berends, M. Eckert (Eds.), The Early Solvay Councils and the Advent of the Quantum Era, *European Physical Journal Special Topics*, vol. 224, n° 10, 2015, 2113-2125.

13. Letter from Einstein to Besso, October 21, 1911, CPAE, vol. 5, doc.296.

14. See Max Jammer, The Conceptual Development of Quantum Mechanics, in *The History of Modern Physics 1800-1950*, American Institute of Physics, vol. 12, Tomash, 1989, 49.

15. An expression due to John L. Heilbron (2013), *op. cit.*, 2.

16. Some sources, to our knowledge, have never been used. This is particularly the case with a "Register" containing the many notes written by E. Solvay during the years1910-1921. This Register, kept at the Université Libre de Bruxelles, is part of the Archives of the Belgian Chemical Society (Fonds SCB, S.a.b.ULB). See Annex 2, Sources relating to Ernest Solvay's work.

17. Solvay therefore needed help from professional physicists.

18. Solvay wanted to find his International Institutes for a period of 30 years.

19. This desire was a response to his conviction that a fundamental law, related to gravity, governed phenomena in fields as varied as physics, chemistry, physiology and sociology. We will come back to Solvay's Gravito-Materialitic approach later.

20. Emile Tassel, *Notes sur les travaux poursuivis par Ernest Solvay de 1857 à 1914*, Brussels, G. Bothy, 1920 (see Annex 2). The author seems to have wanted to remain anonymous, but it is clear that it could only be Tassel.

21. Ernest Solvay, *Sur l'établissement des principes fondamentaux de la gravito-matérialitique*, Bruxelles, G. Bothy, October 1911.

22. Archives of the Belgian Chemical Society, undisclosed source (see Annex 2).

23. The authors' translation of Einstein's words; see also CPAE, vol. 5, doc.270.

24. We may assume that the picture was taken on October 31, 1911, after Planck's presentation of his report (his radiation law is written on the blackboard).

25. Some scientists had taken part in foreign conferences, but these were scholars who spoke several languages.

26. Letter from Lord Rayleigh to Nernst, July 13, 1911, FIS, S.a.b.ULB, doc.1731.

27. Louis D'Or and Anne-Marie Wirtz-Cordier, *Ernest Solvay*, Académie Royale de Belgique, Mémoires de la Classe des sciences, 2ème série, T.XLIV, Fascicule 2, 1981, chap. III, La naissance du procédé Solvay; K. Bertrams, N. Coupain, E. Homburg, *Solvay: History of a Multinational Family Firm*, Cambridge University Press, 2015.

28. L. D'or and A.-M. Wirtz-Cordier, *op. cit.*, 47.

29. *Ibidem*, 48.

30. E. Tassel, *op. cit.*, 29.

31. Solvay is representative of what is called in Belgium "a liberal", i.e., not a socialist but someone who believes in social progress. Concerned with the well-being of the workers in his factories, he provided them with benefits which far anticipated legal provisions. In 1913 he offered 1 million francs to the Belgian Worker's Party for the creation of an Institute of Workers Education at Brussels' "Maison du Peuple". See L. D'Or and A.-M. Wirtz-Cordier, *op. cit.*, 33 and 40.

32. Jean-François Crombois, *L'Univers de la sociologie en Belgique de 1900 à 1940*, Bruxelles, Editions de l'Université Libre de Bruxelles, 1994, 32-33.

33. Stas strongly opposed Solvay's energetic concepts, see L. D'Or and A.-M. Wirtz-Cordier, *op. cit.*, 48.

34. This essentially German doctrine was a reaction to the efforts of physicists to develop a purely "mechanical" view of the universe.

35. Wilhelm Ostwald, *Lebenslinien*, Berlin, 1927, 322.

36. Solvay had added a business school to his Leopold Park foundations.

37. Jacques Bolle, *Solvay, l'invention, l'homme, l'entreprise industrielle*, Bruxelles, Weissenbruch, 1963, 196.

38. See Annex 3: Solvay's Gravito-Materialitic program.

39. Letter from Solvay to Ostwald, May 17, 1913, FIS, S.a.b.ULB.

40. See Annex 3: Solvay's Gravito-Materialitic program.

41. Mayer's name figures prominently at Solvay's Physiology Institute. See J.-J. Heirwegh and M. Peeters, Le tableau de chevet d'Ernest Solvay, in A. Despy-Meyer and D. Devriese (Eds.), *Ernest Solvay et son temps*, Bruxelles, Archives de l'Université Libre de Bruxelles, 1997, 185.

42. Attempts to apply Fresnel's method had been made in France and England, but they had proved unsuccessful; see L. D'Or and A.-M. Wirtz-Cordier, *op. cit.*, 14.

43. Henri Poincaré, *La valeur de la science*, Paris, Flammarion, 1905, 220.

44. Let us recall that Poincaré considered gravitation as the *"less imperfect physical law"* (*ibidem*, chapter XI), and that he announced a crisis in mathematical physics. Several elements attest to Poincaré's influence on Solvay, in particular the importance given by the industrialist to radium, *"the great revolutionary"*, and to Brownian motion, two phenomena cited by Poincaré as being able to endanger the laws of thermodynamics. See also Solvay's letter to Lorentz of August 22, 1912, FIS, S.a.b.ULB, doc.149, and his Note of October 24, 1912 (undisclosed source, Archives of the Belgian Chemical Society).

45. Ernest Solvay, *Gravitique. De la gravité astronomique de la matière et de ses rapports vers l'énergie. Notes, lettres et discours d'Ernest Solvay*, vol. 1, *Gravitique et Physiologie*, Bruxelles, Lamertin, 1929.

46. Nicolas Coupain, Ernest Solvay's Scientific Networks. From Personal Research to Academic Patronage, in F. Lambert *et al.*, *op. cit.*, 2084.

47. Archives de la Société Chimique de Belgique, 17-22; see Annex 2.

48. E. Tassel, *op. cit.*, 21

49. Augustin Boutaric, *La lumière et les radiations invisibles*, Flammarion, 1925, and John Shipley Rowlinson, *Sir James Dewar, 1942-1923. A Ruthless Chemist*, London, Routledge, 2016.

50. Hendrik A. Lorentz et Edouard Herzen, *Les rapports de l'énergie et de la masse d'après Ernest Solvay*, Comptes-rendus de l'Académie des sciences, séance du 12 novembre 1923, t. 177, 925.

51. A. Boutaric, *op. cit.*, 230.

52. Excerpt from a "Mémoire" deposited by Solvay at Belgium's Royal Academy. in sealed envelope.

53. E. Tassel, *op. cit.*, 47.

54. R. Goldschmidt, *Navigation aérienne. Les aéromobiles*, Paris, Dunod et Pinat, Bruxelles, Ramlot 1911. See François Stockmans, *Goldschmidt (Robert-Benedict)*, Biographie Nationale, Académie Royale de Belgique, t. XLII, 309.

55. E. Picard, *La science moderne et son état actuel*, Paris, Flammarion, 1905, 110.

56. N. Coupain, *op. cit.*, 2082; J. S. Rowlinson, *op. cit.*, 72.

57. We shall see that this system was adopted by Lorentz when ISIP's objectives, and the ways to reach them, were specified. That Solvay was the first to come up with the idea is attested by this statement in a letter to P. Héger of March 26, 1912: "*This subsidy scheme is not new to my mind, and the fact that the Physics Institute adopts it is proof that it is good*".

58. In particular his Memoir written in 1887 with Tassel and entitled "Gravitique. De la gravité astronomique de la matière et de ses rapports avec l'énergie", and his Essay "On a fundamental synthesis of the Universe" deposited at Belgium's Royal Academy in sealed envelope.

59. It is in these terms that Solvay unveiled his research program on May 12, 1912, during his invited speech at the Belgian Chemical Society.

60. Solvay's theorem on the constancy of the cold produced in successive expansions was published in 1895 in the *Comptes-rendus de l'Académie des sciences de Paris*. His improvement of the Coleman process is mentioned by J. S. Rowlinson (*op. cit.*, 98). Solvay succeeded in reaching a temperature of -93 °C, a performance hailed by Kamerlingh Onnes (H. Kamerlingh Onnes, "Ter herdenking Ernest Solvay", *De Ingenieur* 37, 1922, 554).

61. These celebrations have been reported in the *Proceedings of the Meetings of the Members of the Royal Institution of Great-Britain*, vol. 16, 1899, 197-218. A talk by Lord Rayleigh was followed on June 6 by the appointment of 26 honorary members. Among them: Arrhenius, Becquerel, Nernst, Solvay and Ostwald (absent). A garden party was organized the next day by Dr. Mond and his wife.

62. *Actes du Jubilé de 1909*, University of Geneva, 1910, 87.

63. For a detailed report on this first phase, see L. Rosenfeld, *op. cit.*, or T. S. Kuhn, *Black-Body Theory and the Quantum Discontinuity, 1894-1912*, University of Chicago Press, 1978, 114-232.

64. See Annex 4, The Black-Body Problem.

65. Lord Rayleigh, *Philosophical Magazine* 49, 1900, 539, and *Nature* 72, 1905, 54 and 243. See also J. H. Jeans' paper, *Philosophical Magazine* (6) 10, 1905, 91.

66. This constant would later be called "Boltzmann constant".

67. Planck had made use of a model in which oscillating electric charges were supposed to realize thermal equilibrium between radiation and matter. He had been forced to admit that the energy of these oscillators should be equal to an integer multiple of a fundamental unity, proportional to the oscillation frequency.

68. Lorentz had completed Maxwell's electromagnetic theory with a theory founded on the existence of charged particles (later called electrons). This theory accounted for the interactions between radiation and matter. The development of Lorentz's theory of electrons started in 1877 with a hypothesis according to which matter consisted of particles, part of which were carrying electric charge. A charged particle in motion with respect to the ether (assumed to be at rest) could be assimilated to a current element. The actions of the electromagnetic field on the particle and the latter's reaction on the field were the only links between matter and the ether. See P. Zeeman and A. D. Fokker, *H. A. Lorentz Collected Papers*, Den Haag, 1934-1939, vol. 2, 80.

69. P. Zeeman and A. D. Fokker, *op. cit.*, vol. 3, 155.

70. T. Kuhn, *op. cit.*, 182.

71. *Ibidem*, 138 and 152. Ehrenfest stayed in Leiden as a student in 1903 and attended Lorentz's lectures (M. Klein, *Paul Ehrenfest. The Making of a Theoretical Physicist*, Amsterdam, North Holland, 1970, 217-234).

72. P. Zeeman and A. D. Fokker, *op. cit.*, vol. 7, 317. We shall see that the question of the necessity of the quantum hypothesis would be resolved in 1912 by Poincaré, following the discussions that had taken place at the first Physics Council.

73. CPAE, vol. 5, doc. 149.

74. Letter from Einstein to Lorentz of March 30, 1909, CPAE, vol. 5, doc. 146.

75. Letter from Lorentz to Einstein of May 6, 1909, CPAE, vol. 5, doc. 153

76. This paper of March 1905 was the first to be published by Einstein during his "miraculous year".

77. This effect, discovered by Hertz in 1887, had been studied experimentally by Philipp Lenard.

78. See Siegfried Grundmann, *Einstein's Akte*, Berlin, 1998, 25-26.

79. Robert Millikan, *Physical Review* 4, 1914, 73 and *Physical Review* 7, 1916, 355.

80. Lorentz's project appears in his research program at the Teyler Foundation for the year 1912-1913, see the *Jaarverslagen van het Teyler Natuurkundig Laboratorium*, Archief Teyler Stichting, Haarlem. We shall come back to this point in a next section.

81. See Lorentz's first letter to Einstein (of May 6, 1909), CPAE, vol. 5, doc. 153.

82. M. Klein, 1965, *op. cit.*, 173-180.

83. According to this theory the thermal properties of the solid are determined by the oscillations of its atoms with respect to their equilibrium position.

84. The specific heat of a body is the amount of heat that must be supplied to increase its temperature by one degree.

85. Reformulated by Planck, the heat theorem would become known as the "third law" of thermodynamics.

86. M. Thiesen, Verhandlungen der Deutsche Physikalische Gesellschaft, *Deutsche Physikalische Gesellschaft*, December 18, 1908.

87. Walther Nernst, *Sitzungsberichte der Preußischen Akademie*, 1909, 247.

88. Walther Nernst, *Theoretische Chemie vom Standpunkte der Avogadroschen Regel und der Thermodynamik*, Stuttgart, 1909, 700.

89. A. Eucken, *Physikalische Zeitschrift* 10, 1909, 586.

90. See T. Kuhn, *op. cit.*, 215.

91. Diana Kormos Barkan, 1993, *op. cit.*, 62. The authors wish to express their gratitude to Diana K. Buchwald for drawing their attention to this letter and for providing them with a copy.

92. Letter from Einstein to Laub of March 16, 1910, CPAE, vol. 5, doc.199.

93. Letter from Einstein to Laub of May 17, 1909, CPAE, vol. 5, doc. 160.

94. The famous fluctuation formulas that Einstein managed to derive in 1909 from Planck's radiation law, considered as an experimental fact.

95. Letter from Einstein to Stark of July 31, 1909, CPAE, vol. 5, doc.172.

96. Letter from Einstein to Sommerfeld of July 7, 1910, CPAE, vol. 5, doc. 211.

97. Letter from Einstein to Laub of August 27, 1910, CPAE, vol. 5, doc.224.

98. A. Pais, *Subtle is the Lord...*, Oxford University Press, 1982, 357-359.

99. Walther Nernst, Sur les chaleurs spécifiques aux basses températures et le développement de la thermodynamique, *Journal de Physique* 4, t. IX, 1910, 721-749.

100. Walther Nernst, *Hauptversammlung der Deutschen Bunsen Gesellschaft für angewandte physikalische Chemie*, Giessen, May 5-8, 1910.

101. Walther Nernst, Sur la détermination des affinités chimiques à partir de données thermiques, *Journal de Chimie physique*, March 8, 1910, 267.

102. M. Jammer, *op. cit.*, 47. We shall see that Nernst's attitude changed considerably over the following months.

103. Elisabeth Crawford, *The Beginnings of the Nobel Institution. The Science Prizes, 1901-1915*, Cambridge University Press, 1987, 133.

104. See Annex 5, Planck's "missed" Nobel Prize.

105. Anne J. Kox, Hendrik Antoon Lorentz's Struggle with Quantum Theory, *Archive for History of Exact Sciences*, vol. 67, 2, 2013, 152-157.

106. E. Crawford, 1987, *op. cit.*, 120-121.

107. F. Lindemann, *Physikalische Zeitschrift* 11, 1910, 609.

108. Einstein had pointed out that in some cases this characteristic atomic frequency could be determined from optical data; he had shown that the value derived from the specific heat of the solid agreed with that obtained from its infra-red absorption spectrum (method of "residual rays").

109. It is likely that Nernst compared the situation with that of chemistry in 1860, and that it reminded him of the Karlsruhe Congress (he mentioned this fact in his opening speech of the first Solvay Council; see section 2.8).

110. E. Crawford 1987, *op. cit.*, 99.

111. D. Kormos Barkan, *Walther Nernst and the Transition to Modern Physical Theory*, Cambridge University Press, 1999, 220.

112. E. Crawford, 1987, *op. cit.*, 127-128 and 257-258.

113. *Ibidem*, 223.

114. Françoise Levie, *L'Homme qui voulait classer le monde*, Bruxelles, Impressions nouvelles, 2006, 119.

115. W. Ostwald, 1927, *op. cit.*, 278.

116. It was during this Congress that it was decided, in homage to Marie Curie, that the name "Curie" would be given to the amount of radium emanations in equilibrium with one gram of radium. See Arthur Eve, "Rutherford", Cambridge University Press, 1939, 192. One of the authors (F.L.) wishes to thank Professor Mary Fowler, Master of Darwin College and descendant of Ernest Rutherford, for providing him with a copy of the book.

117. Robert Goldschmidt was the son-in-law of Franz Philippson, an influential banker and a partner of the Solvay Company. It is likely that Philippson created an initial contact between Goldschmidt and Ernest Solvay.

118. F. Stockmans, *op. cit.* This paper contains many details about Goldschmidt's life and activities.

119. We may assume that Goldschmidt's activity in this field was linked to Philippson's interest in wireless telegraphy; see René Brion and Jean-Louis Moreau, *Franz Philippson. Aux origines de la Banque Degroof*, Bruxelles, Didier Devillez, 2016, 147. Goldschmidt created in 1913 the International TSF Commission, the future URSI (IURS).

120. F. Levie, *op. cit.*, 107.

121. This airship flying over the Grand-Place in Brussels would be chosen as the emblem of the International Exhibition of 1910; see Serge Jaumain and Wanda Balcers, *Bruxelles 1910. De l'Exposition à l'Université*, Bruxelles, Editions Racine, 2010.

122. F. Stockmans, *op. cit.*, 303.

123. W. Nernst, 1910, *Sur les chaleurs spécifiques aux basses températures...*, *op. cit.*, 721.

124. Honorary President: Ernest Solvay, President: Lord Rayleigh, Secretaries: Goldschmidt and a younger person. Members: Einstein, Knudsen, Hasenöhrl, Lorentz, Langevin, Perrin, van der Waals, Larmor, Jeans, Schuster, J.J. Thomson, Rutherford, Nernst, Planck, Wien, Röntgen, Seeliger.

125. Planck's original letter has been lost, but a copy has been made by Jean Pelseneer in an unpublished manuscript: *Historique des Instituts internationaux de physique et de chimie Solvay, depuis leur fondation jusqu'à la Deuxième Guerre Mondiale*, FIS, S.a.b.ULB, doc.1720. A microfilm reproduction of the document has been included in the *Archive for History of Quantum Physics*, Library of the American Philosophical Society. A copy of Pelseneer's "Historique" has been kept in Brussels by be the Service des Archives et des Bibliothèqques de l'ULB.

126. The precise date of the encounter is not known.

127. Our translation in English of the original letter, FIS, S.a.b.ULB, doc.1688a.

128. It should be noted that in 1910 neither Lorentz (in his *Wolfskehl* lectures), nor Planck (in the third edition of his *Thermodynamics*, published in 1911), made any reference to Einstein's theory of specific heats.

129. FIS, S.a.b.ULB, doc. 1688b.

130. Max Planck *et al.*, *Vorträge über die Kinetische Theorie der Materie und der Elektrizität*, Leipzig, Teubner, 1914. See also A. Schirrmacher, 2015, *op. cit.*, 2116.

131. Karl Darrow, Secretary of the American Physical Society, made contact with two members of previous Solvay Councils, Léon Brillouin and Wolfgang Pauli; see S. S. Schweber, A Short History of Shelter Island I, in "Proceedings of the 1983 Shelter Island Conference on Quantum Field Theory and the Fundamental problems of Physics", Eds. R. Jackiw, N.N. Khuri, S. Weinberg and E. Witten, 303.

132. P. Langevin et M. de Broglie (Eds.), *La théorie du rayonnement et les quanta. Rapports et discussions de la réunion qui s'est tenue à Bruxelles du 30 octobre au 3 novembre 1911 sous les auspices de M. E. Solvay*, Paris, Gauthier-Villars, 1912.

133. The list of participants was an important element, enabling the invitees to decide whether, or not, they should accept to attend the Council. This would notably be the case with Rutherford, who expressed hesitation in a letter to Nernst of June 17, 1911, FIS, S.a.b.ULB, doc.721.

134. This condition would reappear in a less restrictive form in the official invitation letter, signed by Solvay.

135. D. Kormos Barkan, 1999, *op. cit.*, 149 and 157.

136. Archives of the Belgian Chemical Society, partially diffused source (see Annex 2).

137. Remember that Solvay had obtained this rule in 1858.

138. See his opening speech of the Council, section 2.8.

139. Solvay's rejection of kinetic theory may easily be understood when one recalls the answer reported by Albert Ducroc to the question: "What interest did the kinetic theory of gases offer in 1902, date of the publication of Boltzmann's work?

"For the philosopher it was devoid of any poetry, and the industrialist considered it useless because it didn't allow him to improve the performance of his machines". See A. Ducroc, *Le roman de la matière*, Paris, Union Générale d'Editions, 1963, 30.

140. We may assume that his father's position played a role in Herzen's appointment to a laboratory of the Solvay Company (he would become the laboratory's director a few years later).

141. This edition, *Traité pratique d'électrochimie*, by Georges Hostelet and Richard Lorenz, was published in 1905 (Paris, Gauthier-Villars). For the German edition, see R. Lorenz, *Elekrochemisches Praktikum*, Göttingen, Vandenhoeck und Ruprecht, 1901.

142. Great resistant during the War, alongside nurse Edith Cavell, Hostelet was imprisoned in Germany. He later devoted himself to sociology. The authors thank the historian Hippolyte Bailly for having provided them with the above information, taken from his Memoir of 2013 (University of Louvain): "Resister dans le réseau Cavell: histoire, mémoire, trajectoires de vie, Georges Hostelet (1875-1960). They also thank Pierre Verhas, Hostelet's grandson, for providing them with a photo of Hostelet, taken in Cairo in the early 1920's.

143. See Solvay's letters to Hostelet of the years 1910-1911, and Solvay's typed Note of October 5, 1910, Archives of the Belgian Chemical Society, undisclosed source 16, 17 (see Annex 2).

144. Letter from Nernst to Solvay of November 27, 1910, FIS, S.a.b.ULB, doc.1689b.

145. *Ibidem*, doc.1690a.

146. CPAE, vol. 5, doc.230.

147. A. J. Kox, Einstein, Specific Heats and Residual Rays: The History of a Retracted Paper, in A. J. Kox and D. M. Siegel (Eds.), No Truth Except in the Details, *Boston Studies in the Philosophy of Science*, Kluwer, 1995, 249.

148. *Proceedings of the Prussian Academy of Science*, January 6, 1911. See also T. Kuhn, *op. cit.*, 214.

149. This formula contained two terms similar to Einstein's: one was term which appeared in the latter's formula (with the usual proper frequency), the other was a term which involved half of that frequency. Nernst associated these two terms with Rubens' residual ray-results; see A. J. Kox, 1995, *op. cit.*, 250.

150. Letter from Solvay to Goldschmidt of March 12, 1911, FIS, S.a.b.ULB, doc.1691.

151. Nernst' list only contained two French physicists: Langevin and Perrin (see note 124).

152. Martha Cecilia Bustamante, *Rayonnement et quanta en France, 1900-1914*, Firenze, S. Olschki (Edit.), 2003, 65.

153. Solvay's Note of September 15, 1910, Archives of the Belgian Chemical Society, undisclosed source (see Annex 2).

154. See Solvay's letter to Brillouin of March 5, 1913, FIS, S.a.b.ULB, doc.1123.

155. Patrick Juignet, *L'idée de matière*, Philosophie, science et société (en ligne), 2017.

156. Letter from Herzen to Solvay of March 13, 1911, FIS, S.a.b.ULB, doc. 1692.

157. We saw (Chapter 1) that it was decided to offer each guest a sum of 1,000 francs for "travel expenses", with the understanding that subsistence costs in Brussels would be borne by Solvay.

158. Herzen must have noticed that the sum for travel expenses, indicated in his previous letter, should be increased. We will see that Nernst suggested to reduce the amount to 1,000 francs.

159. M. Klein, Thermodynamics in Planck's Work. History of Physics, in S. R. Weart and M. Phillips (Eds.), *Readings from Physics Today*, The American Institute of Physics, New york, 1985, 300.

160. In fact, only Lorentz, Planck and Knudsen had been consulted and had reacted positively. Nernst apparently displayed frank optimism in reaction to Planck's reservations (see his letter to Nernst of June 11, 1910).

161. Lorentz's interest in the theorem is attested by the Faculty of Leiden's decision to put the following question to the competition (May 1, 1911): "*The Faculty wishes a theoretical presentation on the consequences of Nernst's theorem; this presentation will include a comparison of the predictions of the theorem with experimental data; it will examine the extent to which the theorem can be deduced from general principles, and will clarify its links with the current concepts of the theory of heat*".

162. FIS, S.a.b.ULB, doc.1696b.

163. We will see that Kamerlingh Onnes had the opportunity to present to the Council his recent graphs showing the collapse of the electrical resistance of mercury at a temperature close to 4° Kelvin.

164. Their formulation in the official invitation may be slightly different from what follows. More important deviations will appear in the Council's program: changes in the titles of some reports, presentations that hadn't been scheduled.

165. FIS, S.a.b.ULB, doc.1695.

166. Archives of the Belgian Chemical Society, undisclosed source (see Annex 2).

167. FIS, S.a.b.ULB, doc.1695.

168. *Ibidem*, doc.1697.

169. Regine Zott, *Wilhelm Ostwald und Walther Nernst in ihren Briefen*, Berlin, 1996, 186.

170. See note 62. The Geneva Faculty of Science proposed 20 honorary doctorates. The list of recipients comprised the following names: Einstein, Guillaume, Haller, Ostwald, Schuster, Solvay, Voigt. Marie Curie was to receive a doctorate in medicine.

171. W. Ostwald, *Annalen der Naturphilosophie* 6, 1907, 480.

172. *Berichte der Deutschen Chemischen Gesellschaft* 40, 1907, 4304 and 5031. The proposal to elect Solvay honorary member was signed by 41 members. Among them: Fischer, Nernst, Ostwald and van 't Hoff. Nernst was elected president of the Gesellschaft for the year 1908.

173. F. Levie, *op. cit.*, 119.

174. *Actes du Congrès mondial des Associations internationales: Documents préliminaires, Rapports, Procès-verbaux*, publiés par l'Office des Associations internationales, Bruxelles, Hayez, 1911, 13-18.

175. Ido, a language derived from Esperanto, see Michael D. Gordin, *Scientific Babel*, Chicago, 2015.

176. Dony had worked in Göttingen under Nernst's direction, and had obtained in 1900 the title of Doctor in Chemical Science. He became attached to Solvay in 1901 as manager of a laboratory at Park Léopold's Physiology Institute.

177. This Institute had been created by Edmund Knowles Muspratt, the son of the industrialist James Muspratt. It had been inaugurated in October 1906, in the presence of Ostwald and Sir William Ramsay.

178. *Donnan Papers*, UCL Special Collections, University College London, GB 0103.

179. This Electrochemical Institute received a donation of 100,000 francs from the Solvay Company. See Françoise Birck, *Des ingénieurs pour la Lorraine*, Editions Serpenoise, 1998, 159. We shall come back to this point in the second part of the book, in connection with Albin Haller's role in the creation of ISIC.

180. Ostwald's ambitious project provided for a close link between the Institute and the IACS. The sketch would be developed in 1912 in a Memoir: "Denkschrift über die Gründlung eines internationalen Institutes für Chemie".

181. Letter from Solvay to Ostwald of June 23, 1911, FIS, S.a.b.ULB, dossier IICS.

182. The letters from Solvay to Ostwald (of June 23 and July 7) are part of the Wilhelm Ostwald Archiv, Berlin Brandenburg Akademie der Wissenschaften (BBAW) The authors are grateful to Diana K. Buchwald for drawing their attention to these letters and for providing them with copies. They also wish to thank Dr. Vera Encke, director of the Archives of the BBAW.

183. Message from Ostwald to Dony of July 13, 1911, FIS, S.a.b.ULB, dossier IICS.

184. *Ibidem*, dossier IICS.

185. FIS, S.a.b.ULB, doc.1712

186. *Ibidem*, doc.1736.

187. *Ibidem*, doc.1737.

188. *Ibidem*, doc.1738.

189. See Annex 6, The second Moroccan crisis and the Caillaux affair.

190. Letter from Nernst to Goldschmidt (back from his trip), FIS, S.a.b.ULB, doc.1739.

191. Planck was struck by Solvay's description of the revolution of a body around a fixed point. We cannot help seeing it as related to the mode adopted by Niels Bohr to describe the quantized states of the atom.

192. Four reports presented at the Council – those of Warburg, Rubens, Perrin and Langevin – do not appear on this list. The first two hadn't been communicated on time, the last one hadn't been scheduled. It seems that Herzen didn't realize the importance of the link between Perrin's results and Planck's theory (we will see that he contented himself with evoking in his biographical notes "*Perrin's fine research on Brownian motion*").

193. Herzen refers the reader to page 62 of Solvay's preliminary study, see note 21.

194. Théophile de Donder observed much later the existence of a link between the "small cubical elements" imagined by Solvay and the cells of modern statistical quantum theories, the volume of which is determined by Planck's constant. See his letter to Lefébure of September 22, 1937, FIS, S.a.b.ULB, doc.3470.

195. *Ibidem*, docs. 1758, 1769, 1774. A notebook with the individual photos can be found in Leiden (Kamerlingh Onnes Archive of Rijskmuseum Boerhaave, RMB).

196. Lord Rayleigh did announce his intention to send a letter to the Council "out of respect for his colleagues".

197. See John L. Heilbron, *British Participation in the First Solvay Councils*, in F. Lambert et al (Eds), *op. cit.*, 2041-2055.

198. See "Le XXVe anniversaire de l'Institut de Sociologie Solvay", *Revue de L'Institut de sociologie* 4, Bruxelles, Imprimerie scientifique et littéraire, 1927, 13. A summary of the conferences of the British MP's (presented before 1910) had been published in a volume entitled "La politique de réforme sociale en Angleterre". Let us recall that in 1911 the United Kingdom was shaken by massive strikes (during the Great Labour Unrest), and that H. H. Asquith, a member of the Eighty Club, had launched a campaign to reduce the power of the House of Lords, controlled by the Conservative Party.

199. Goldschmidt had told Solvay that Dony's appointment as "deputy manager" would please Nernst (FIS, S.a.b.ULB, doc.1742). It seems that Nernst profited of such possibility to give the task to Lindemann instead.

200. Nernst insisted that a reminder be sent to van der Waals, but the Dutch physicist mentioned health problems and persisted in his refusal to come to Brussels.

201. Alfred Stock, *Internationales Chemiker-Kongres 1860*, Berlin, 1933, contribution to the Hauptversammlung der Deutschen Bunsen-Gesellschaft.

202. The idea of accommodating all guests in the same hotel became a rule for the following Councils. We will see that this option gave rise to the famous Bohr-Einstein debates of 1927 and 1930, which took place on the sidelines of the official Council sessions. However, it should be noted that Hotel Métropole would no longer be chosen after the first Solvay meeting (the hotels for the later years will be indicated).

203. M. de Broglie, *Registre contenant des pièces manuscrites concernant les premiers Congrès Solvay*, offert à l'Académie des sciences de Paris en la séance du 19 décembre 1951 (2 volumes), Paris, Archives de l'Institut de France.

204. Titles of these reports: *Experimentele Prüfung der Planckschen Formel für Hohlraumstrahlung*, von E. Warburg; *Prüfung der Planckschen Strahlungsgleichung im langwelligen Spectralbereich*, von H. Rubens; *Die Gesetze der Wärmestrahlung und die Hypothese der elementaren Wirkungsquanten*, von M. Planck; *Die kinetische Theorie der ideale Gase und die Versuchsresultate*, von M. Knudsen; *Anwendung der Quantentheorie auf eine Reihe physikalisch-chemischer Probleme*, von W. Nernst; *Die Bedeutung des Wirkungsquantums für unperiodische Molekularprozesse*, von A. Sommerfeld; *Zum gegenwärtigen Stande des Problems der spezifischen Wärme*, von A. Einstein.

205. Lord Rayleigh's letter is not to be considered as a Solvay report.

206. Kamerlingh Onnes presented a graph, obtained just a few days before the Council's opening. It showed the spectacular collapse of the electrical resistance of mercury at about 4° Kelvin.

207. Only one question, by Langevin, was reported in the official volume. It gave rise to only one comment (from Kamerlingh Onnes).

208. M. Klein, 1985, *op. cit.*, 300-302.

209. A first draft of the Council's report was submitted to the contributors on December 23, 1911. See the "Solvay file", Fonds Brillouin, Archives du Collège de France. The authors thank Mrs. Anne Chatellier for giving them access to this file.

210. Poincaré took advantage of this possibility to rectify (in a footnote) the point of view about energy quanta which he had defended during the Council's general discussion.

211. See Solvay's closing speech in a next section.

212. We know that Goldschmidt suggested Solvay to create an International Physics Institute (FIS, S.a.b.ULB, doc.24a). We will see (section 3.2) that a subsidy request was introduced by the Russian physicist Piotr N. Lebedew, more than two months before ISIP's foundation (proof that the news about the granting of Solvay subsidies reached Russia in January 1912, following an indiscretion of a Council member who had taken part in the private meeting with Solvay on November 3, 1911).

213. The authors thank their colleague Michael Eckert for providing them with an excerpt of Sommerfeld's letter. They are also grateful to Mrs. Monika Baier, grand daughter of Sommerfeld, for sending them a copy of the letter.

214. See Annex 6: The second Moroccan crisis and the Caillaux affair.

215. See Brillouin's letter to Tassel of December 26, 1918, FIS, S.a.b.ULB, doc.673.

216. Letter of November 28, 1911; Archive Wander de Haas (Lorentz's son-in-law), RMB, Leiden, 8-9.

217. M. Brillouin, H.A. Lorentz en France et en Belgique, *Physica* 6, 1925, 30.

218. Only Solvay's speech has been reported in the official volume. Three closing speeches – by Lorentz, Rutherford and Goldschmidt – can be found in de Broglie's report (see note 203).

219. In his answer to the invitation to attend the Council (CPAE, vol. 5, doc.270), Einstein had drawn Nernst's attention to the fact that it wouldn't be possible to maintain quantum theory, in its actual form, if the prediction of an exponential decrease of the thermal conductivity in the vicinity of absolute zero was invalidated by experience. Unfortunately, this prediction had been contradicted by Nernst's and Eucken's recent results. See P. Langevin and M. de Broglie (Eds.), *op. cit.*, 421.

220. It should be noted that most of the Council members spoke of energy quanta, whereas Planck's new concept implied the need to distinguish between the emission and the absorption of radiation. Only Einstein was prepared to associate quanta with "particles of light". For the other members the quantum concept only intervened in the processes of emission and absorption of radiation.

221. See Einstein's report in Langevin and de Broglie (Eds.), *op. cit.*, 428.

222. Einstein referred to the conclusion of a note published on January 2, 1911: "Remark on a fundamental difficulty in theoretical physics", CPAE, vol. 3, *The Swiss Years: Writings 1909-1911*, doc.16.

223. Einstein's report in Langevin and de Broglie (Eds.), *op. cit.*, 429.

224. *Ibidem*, 431.

225. Letter from Einstein to Besso, CPAE, vol. 5, doc.331.

226. Letter from Einstein to Besso of May 13, 1911, CPAE, vol. 5, doc.7.

227. Letter from Einstein to Zangger of November 20, 1911, CPAE, vol. 5, doc.308

228. Letter from Einstein to Besso of February 4, 1912, CPAE, vol. 5, doc.354.

229. See E. Crawford, J. L. Heilbron and R. Ulrich, The Nobel Population 1901-1937. A Census of the Nominations and Nominees for the Prizes in Physics and Chemistry, *Berkeley Papers in History of Science* 11, 1987; see also A. Pais, 1982, *op. cit.*, 153.

230. Speaking in Düsseldorf in 1898, Wien had put his focus on the ether-problem. It seems that his lecture had been a source of inspiration for Einstein. See Walter Isaacson: *Einstein, His Life and the Universe*, London, 2007, 48.

231. A. Eve, *op. cit.*, 193.

232. P. Langevin and M. de Broglie, *op. cit.*, 451.

233. H. Poincaré, *Sur la théorie des quanta*, Comptes-rendus de l'Académie des Sciences, Paris, 1911, 1103-1108, et *Journal de physique théorique et appliquée*, Sér.5, 2, 1912, 5-34.

234. Letter from Einstein to Zangger of November 15, 1911, CPAE, vol. 5, doc.305.

235. Letter from Brillouin to Lorentz of January 30, 1912, Haarlem, Noord-Hollands Archief (NHA), Archief H. A. Lorentz (364), inv. num.11.

236. R. MacCormmach, Henri Poincaré and Quantum Theory, *Isis* 58, 1967, 51-53. Jeans had tried to explain thermal radiation data without appealing to quantum theory.

237. E. Rutherford, *Philosophical Magazine*, Series 6, vol. 21, May 1911, 669.

238. We will see that Rutherford defended his nuclear hypothesis in 1913, during the discussion of J. J. Thomson's report to the second Solvay Council.

239. Haas had studied a model of "J. J. Thomson type" with an atomic radius that depended on Planck's constant h.

240. N. Bohr, The Solvay Meetings and the Development of Quantum Physics, in R. Stoops (Ed.), *La théorie quantique des champs* (Douzième Conseil de Physique), New York, Interscience, 1962, 17.

241. L. de Broglie, Vue d'ensemble sur mes travaux scientifiques, in *Louis de Broglie. Physicien et penseur*, Paris, Albin Michel, 1953, 458.

242. D. Kormos Barkan, 1999, *op. cit.*, 181.

243. E. Gubin, Dans la presse. Marie Curie et les premiers Conseils Solvay, in P. Marage and G. Wallenborn (Eds.), *op. cit.*, 59.

244. We borrowed most of what follows from the book of Françoise Giroud "Une femme honorable. Marie Curie, une vie", Paris, Fayard, 1981, 213-245.

245. E. Crawford, 1987, *op. cit.*, 200 and 207.

246. E. Gubin, *op. cit.*, 59.

247. Letter from Rutherford to Meyer of January 22, 1912, see A. Eve, *op. cit.*, 211.

248. Letter from Einstein to Zangger of November 7, 1911, CPAE, vol. 5, doc.303.

249. Translation in English of a letter from Einstein to Marie Curie of November 23, 1911, Albert Einstein Archive (AEA, 84162), courtesy of the Hebraic University of Jerusalem.

250. Letter from Brillouin to Lorentz of January 29, 1912, NHA, 364, inv. num.11. We will come back to this letter in the second part of the book (section 4.2).

251. See Françoise Giroud, *op. cit.* 241.

252. E. Solvay, *Gravitique, De la gravité astronomique de la matière et de ses rapports avec l'énergie*, see Annex 2.

253. The authors thank Nicolas Coupain for providing them with this information.

254. Correspondence of Ernest Solvay, Archive of the Solvay Family. The authors thank Marina Solvay for providing them with copies of some letters. See also, J. Bolle, *op. cit.*, p. 42.

255. It is likely that Nernst was inspired by the *Faraday Discussions*, which aimed at shedding light on specific questions in physical chemistry by giving priority to the discussion of reports. The authors would like to thank Brigitte Van Tiggelen for drawing their attention to this point.

256. A. Einstein, in P. Langevin and M. de Broglie, *op. cit.*, 429.

257. Letter from Einstein to Zangger, CPAE, vol. 5, doc.305.

258. CPAE, vol. 5, doc.277. Julius had attended Einstein's lecture, given in Leiden on February 11, 1911.

259. See CPAE, vol. 5, doc.286.

260. Zangger had attended Einstein's lectures at the Zurich's Cantonal University. He also attended some lectures in Paris by Curie and Langevin (CPAE, vol. 5, doc.291).

261. Letter from Einstein to Zangger of September 20, 1911, CPAE, vol. 5, doc.286.

262. Letter from Einstein to Julius of September 22, 1911, CPAE, vol. 5, doc.288.

263. *Ibidem*, doc.297.

264. Letter from Einstein to Besso, CPAE, vol. 5, doc.296.

265. Not surprising, since mutual trust existed between the two men since Einstein lectured in Leiden in February 1911 and the Einstein couple stayed at Lorentz's home.

266. See next section: The imbroglio of Lorentz's succession.

267. See Einstein's letter to Zangger of November 7, 1911, CPAE, vol. 5, doc.303.

268. *Ibidem*, doc.304.

269. *Ibidem*, doc.305.

270. *Ibidem*, doc.306.

271. *Ibidem*, doc.355.

272. Ronald Clark, *Albert Einstein, The Life and Times*, London, 1973, 167.

273. *Ibidem*, 152. Let us recall that Weiss was an Frenchman, that he met Zangger in Karlsruhe in September 1911, and that he expressed his admiration for Einstein (see CPAE, vol. 5, doc.291).

274. *Ibidem*, 153.

275. Letter from Poincaré to Weiss of November 1911, see R. Debever, Einstein et la Belgique, Compte-rendu de l'Exposition du 16 mai au 19 juin 1979, Académie Royale des sciences et lettres et des beaux-arts de Belgique. See also M. Paty, The Scientific Reception of Relativity in France, *British Society for the Philosophy of Science*, vol. 103, 1987.

276. See Einstein's letter to Zangger of June 1912, CPAE, vol. 5, doc.406.

277. It was imperative for Lorentz to preserve his pension rights. This meant that his leave for Haarlem couldn't take place before the fall of 1912, and that it might even be scheduled for the spring of 1914.

278. It is likely that Lorentz wanted to know better the man who could become his successor in Leiden. Before Einstein, only Planck, colleague and friend of Lorentz, had been invited to stay at the Dutchman's home.

279. See Lorentz's letter to Einstein of February 13, 1912, CPAE, vol. 5, doc.359.

280. This is one of the reasons that Einstein would invoke in a letter to Julius of November 15 to justify his decision to decline Utrecht's offer.

281. NHA, 763, Van der Waals Archief, inv. num.107. Lorentz refrained from mentioning Solvay's ISIP-project in his official correspondance.

282. Letter from Einstein to Lorentz of November 23, 1911, CPAE, col.5, doc.313.

283. This indication confirmed the news revealed by Debije, Einstein's successor at the Zurich's Cantonal University, who was now applying for the position in Utrecht.

284. Letter from Lorentz to Einstein of December 8, 1911, CPAE, vol. 5, doc.318.

285. Letter from Einstein to Lorentz of December 12, 1911, CPAE, vol. 5, doc.322.

286. Universiteitsbibliotheek Leiden, AFA FA12, notulen faculteit.

287. Letter from Lorentz to Einstein of February 13, CPAE, vol. 5, doc.359.

288. Letter from Einstein to Lorentz of February 18, 1912, CPAE, vol. 5, doc.360.

289. Letter from Lebedew to Lorentz, of February 9, 1912, ESPCI, Fonds Langevin, doc.L10/232. If the date corresponds to the Gregorian calendar, it means that Lorentz received the letter before Einstein's announcement of his cancellation.

290. Lebedew's request confirms the fact that the idea of an International Physics Institute, granting research subsidies all over the world, was discussed during the "private meeting" with Solvay of November 3, 1911.

291. Sommerfeld's letter to Lorentz of February 25, 1912 has been preserved, see A. J. Kox (Ed.), *The Scientific Correspondence of H. A. Lorentz 1*, New York, 2008, 353.

292. Ehrenfest and his wife, Tatiana Afanassieva, were the authors of a brilliant paper which appeared in the "Encyclopedia of Mathematical Sciences", edited by Sommerfeld.

293. See M. Klein, 1970, *op. cit.*; see also P. Huynen and A. J. Kox, Paul Ehrenfest's Rough Road to Leiden: a Physicist in Search for a Position 1904-1912, *Physics in Perspective* 9, 2007, 186-211.

294. P. Ehrenfest, Welche Züge der Lichtquantenhypothese spielen in der Theorie der Wärmestrahlung eine wesentliche Rolle?, Annalen der Physik 341, 1911, 91.

295. A notebook, kept at the RMB in Leiden, contains Ehrenfest's words "Lazarew Solvay". They seem to indicate that Ehrenfest intended to forward information about ISIP to Dr. Lazarew, Lebedew's closest collaborator.

296. Letter from Lorentz to Ehrenfest, RMB, Archive Ehrenfest, Leiden.

297. Letter from Solvay to Lorentz, FIS, S.a.b.ULB, doc.27.

298. It should be recalled that this eminent Viennese theorist, father of statistical physics, had been Ehrenfest's thesis director.

299. Letter from Ehrenfest to Lorentz, ESPCI, Fonds Langevin, L10/231.

300. Letter from Lorentz to Ehrenfest of May 13, 1912, RMB, Archive Ehrenfest, Leiden.

301. As mentioned before, Lorentz had learned from Sommerfeld that Einstein thought of Ehrenfest for his succession at the German University in Prague.

302. Einstein would keep this title until July 13, 1946.

303. FIS, S.a.b.ULB, doc 132.

304. See section 1.4, note 80. Lorentz had expressed interest in the subject in 1909 during a conference in Utrecht, in which he spoke of Lenard's results. Yet, he made clear at the time that he considered light as a wave phenomenon, see P. Zeeman and A.D. Fokker, *op. cit.*, vol. 7, 374.

305. See Einstein's letter to Hopf of February 20, 1912, CPAE, vol. 5, doc.364, and his letter to Zangger of about the same time, *ibidem*, doc. 368.

306. See Einstein's letter to Besso of December 26, 1911, *ibidem*, doc.331. For a detailed discussion of Einstein's position with respect to Nernst and Rubens, see A. J. Kox, 1995, *op. cit.*, 245-255.

307. Letter from Haber to Einstein of March 8, 1912, CPAE, vol. 5, doc.368.

308. Letter from Einstein to Elsa Löwenthal of April 30, 1912, *ibidem*, doc.389.

309. See Siegfried Grundmann, *The Einstein Dossiers*, Berlin, Springer, 2004, sect. 1.1.

310. *Ibidem*, 5.

311. CPAE, vol. 5, doc.428.

312. Koppel was a banker and founder of several industrial firms. In 1905, he devoted 1 million marks to a foundation which would promote exchanges between German and foreign intellectuals. More recently he had taken part in the creation of the Kaiser Wilhelm Institute for Physical Chemistry and Electro-chemistry. See S. Grundmann, *op. cit.*, 12, and R. Clark, *op. cit.*, 180, for more details.

313. We will see that Einstein would show himself sensitive to the honor of occupying van 't Hoff's chair.

314. Koppel, as we know, agreed to pay the Academy the funds to provide an important supplement to Einstein's salary.

315. Letter from Einstein to Elsa Löwenthal of July 14, 1913, CPAE, vol. 5, doc.451.

316. Einstein would live there separated from his wife and sons, who remained in Zurich.

317. R. Clark, *op. cit.*, 149.

318. Chris Koenig, How Einstein fled the Nazis to an Oxford College, *The Oxford Times*, March 2012.

319. Terms that Lorentz would use in his letter to Solvay of January 4, 1912.

320. Translation in French of the "Denkschrift..." published by Ostwald in 1912, see note 180.

321. Letter already cited in note 319.

322. Ostwald's project was extremely ambitious. The Institute was supposed to prepare a catalog of all chemical substances, open to any interested researcher; it was to produce a chemical register and a database, including the addresses of the authors of chemical publications, recent or old. It should comprise an experimental section where a collection of all chemical substances would be kept in the form of reliable preparations, as well as a section responsible for providing the translation of all articles and for disseminating a universal language, specific to chemistry ("Ido", a new esperanto of which Ostwald had laid the groundwork). In addition, the Institute was to provide housing for its entire staff.

323. Letter from Solvay to Lorentz of January 9, 1912, FIS, S.a.b.ULB, doc.2a.

324. *Ibidem*, doc.2b.

325. Letter from Hostelet to Solvay of January 7, 1912, FIS, S.a.b.ULB, doc. 4.

326. Hostelet's letter tells us that Solvay discussed the project with his collaborators, considering that the Institute would be responsible for perpetuating the Physics Council.

327. This point doesn't appear explicitly in Héger's notes, but it is clear that Solvay's emissary had to include such a clause in the draft statutes.

328. Letter from Lorentz to Solvay of February 2, 1912, FIS, S.a.b.ULB, doc.107.

329. *Ibidem*, doc.14. Solvay's request reveals a concern that he repeatedly expressed in his writings: the need to make physics "more objective".

330. Letters from Brillouin to Lorentz (end of January 1912), NHA, 364, inv. num.11.

331. Brillouin means the Select Committee (provisional Scientific Committee) which met with Solvay on November 3, 1911.

332. Paul Villard, discoverer of the gamma rays.

333. Charles Fabry, specialist of optical cavities, future president of the French Physical Society (1924) and future foreign member of the Royal Society (1931).

334. Jules Macé de Lépinay, co-founder of the Marseille Engineering School, had been Fabry's thesis director. It is in Macé de Lépinay's laboratory that Fabry and Alfred Perot conceived the Fabry-Perot optical interferometer.

335. Paul Sabatier, catalysis specialist in organic chemistry, Nobel Prize Chemistry in 1912 with Victor Grignard.

336. See chapter 7: The Solvay subsidies.

337. Letter from Héger to Lorentz, FIS, S.a.b.ULB, doc.111.

338. Letter from Lorentz to Solvay of March 6, 1912, *ibidem*, doc. 115.

339. Lorentz was not aware of Lebedew's recent death.

340. Letter from Lorentz to Solvay of March 25, 1912, FIS, S.a.b.ULB, doc.21.

341. 1,000 francs would represent today the equivalent of 6,000 euros.

342. FIS, S.a.b.ULB, doc.27a.

343. *Ibidem*, doc.28.

344. *Ibidem*, doc.126.

345. Letter from Solvay to Lorentz of May 4, 1912, *ibidem*, doc.32.

346. Notes written by Solvay during the spring of 1912, Archives of the Belgian Chemical Society, undisclosed source (see Annex 2).

347. See Solvay's request of February 1913, section 6.5.

348. FIS, S.a.b.ULB, doc.19.

349. *Ibidem*, doc.27a.

350. *Ibidem*, doc.89.

351. Reminder: Héger reacted favourably to Lorentz's proposal.

352. FIS, S.a.b.ULB, doc.41.

353. Letter from Solvay to Marie Curie of November 23, 1911, BNF, Gallica, *Pierre et Marie Curie Papiers II*, docs.266-267. The authors thank Nicolas Coupain for providing them with a copy of these letters.

354. Solvay intervened several times in a generous and discrete manner. See his letters to Marie Curie of July 1913 and April-May 1914 (section 5.2).

355. Solvay seems to refer to a personal endeavour: his desire to participate to this *pursuit of truth* in his own way. We will see that he devoted considerable efforts to it.

356. Archives de l'ESPCI, Fonds Langevin, L10/71. Lorentz asked Knudsen to send the circular letter in the following order: Rutherford, Brillouin, Curie, Warburg, Nernst, Kamerlingh Onnes.

357. FIS, S.a.b.ULB, doc.148.

358. Confidential letter from Lorentz to Solvay of August 17, 1912, *ibidem*, doc.41.

359. *Ibidem*, doc.149a.

360. *Ibidem* doc.154.

361. See Annex 7, Royal patronage.

362. Example: *"One must realize by on the spot observation of the indescribable and eminently sad situation in which Russian university education has been plunged by the vandalism of the Minister of Education, to measure the significance of the safeguarding of Lebedew's laboratory, and its tradition, for the future of Russian physics"*.

363. FIS, S.a.b.ULB, doc.38c.

364. Correspondance Rutherford-Knudsen, June 1912, Archives de l'ESPCI, Fonds Langevin, L10/233.

365. A. Pais, *Niels Bohr's Times*, Oxford, Clarendon Press, 1991, 133. See also Jens Rud Nielsen, Memories of Niels Bohr, *Physics Today*, vol. 16, 10, 1963, who tells us that he had tried in 1921 to put a text of Knudsen in agreement with modern physics, and that the latter had answered: "*If you need quantum theory to explain this point, it is as well not to explain anything at all...*".

366. Correspondance Nernst-Knudsen, June 1912, Archives de l'ESPCI, Fonds Langevin.

367. FIS, S.a.b.ULB, doc.1161.

368. *Ibidem*, doc.1147, Solvay's telegram of August 18, 1912. Lorentz was aware that ISIP's Scientific Committee hadn't met yet. He therefore took care to seek Solvay's approval.

369. *Ibidem*, doc.154c.

370. These minutes are kept in Paris, Archives de l'ESPCI, Fonds Langevin.

371. Letter from Laue of August 17, 1912, Archives de l'ESPCI, Fonds Langevin.

372. Sommerfeld, who wasn't member of the ISC, had clearly been informed of the possibility to apply for a Solvay grant. Knowing hat the Committee would support research on Röntgen rays, he transmitted the information to Laue.

373. Letter from Einstein to Hopf of June 12, 1912, CPAE, vol. 5, doc.408. Einstein joined a drawing representing Laue's experiment in Munich. He called *photograph* a picture which was obtained without the use of a camera: it showed the interference of X-rays diffracted by the crystal (Laue used the term "photogram").

374. According to the *Baedeker* of 1911-1914. On pound was worth 25 francs (i.e. about 20 marks).

375. FIS, S.a.b.ULB, doc.1128.

376. Stefan L. Wolff, Physiker im "Krieg der Geister", *Historical Studies in Physical and Biological Sciences* 33, 2003, 337-368.

377. Wood, a specialist of resonance radiation and absorption spectra, was known for having discovered the error of the French physicist René Blondlot, who claimed to have detected a new type of radiation – the s.c. N rays, capable according to him of increasing the luminosity of light of weak intensity – an announcement which had aroused the German Kaiser's interest.

378. The Council's official report, published in Paris (Gauthier-Villars) after the Great War, mentions ten Solvay reports (two of which hadn't been scheduled: one by Marie Curie, the other by Sommerfeld).

379. Thomson based his criticism on the fact that an atomic nucleus should produce deviations of beta-particles, similar to those of alpha-particles, whereas no such deviations had been observed (see Annex 8, Essential points in the Rutherford-Thomson confrontation).

380. This is what emerges from the report of the meeting of the British Association for the Advancement of Science, held at Birmingham in September 1913 (see the article in *Nature* of November 1913, 304-305). Several Council members had taken part in this meeting: Marie Curie, Lorentz, Wood, and four British physicists (Rutherford, J. J. Thomson, Jeans and W. H. Bragg).

381. Their method opened the way to the discovery in 1953 of the structure of DNA by Francis Crick and James Watson.

382. W. H. Bragg and W. L. Bragg, *X Rays and Crystal Structure*, London, G. Bell and Sons, 1918.

383. Lorentz recalls what he had written about the heat theorem in July 1913. See *Chemisch Weekblad* 10, 1913, 621; see also P. Zeeman and A. D. Fokker, *op. cit.*, VI, 318.

384. Statement made by Einstein during the discussion of the Grüneisen report; see *La Structure de la matière, Rapports et discussions du Conseil de physique tenu à Bruxelles du 27 au 31 octobre 1913 sous les auspices de l'Institut international de physique Solvay*, Paris, Gauthier-Villars, 1921.

385. Letter from Einstein to Hopf of November 2, 1913, CPAE, vol. 5, doc.480.

386. *Ibidem*, doc.482.

387. The authors thank Mrs. Monika Baier, grand daughter of Sommerfeld, for providing them with a copy of two letters written by Sommerfeld during the Council, and for authorizing them to reproduce the following excerpts.

388. This intercontinental station had been built by Goldschmidt in Laeken's royal domain. See F. Stockmans, *op. cit.*, 318: "*This station was one of the most powerful known: it included an antenna of 600 meters (supported by four pairs of pylons, one of 120 meters high, the other three of 65 meters) and of a proper wavelength of 3,500 meters. Its commissioning had been laborious because of the large quantity of electric energy, formidable for the time, which had to be domesticated*".

389. Sommerfeld's complement to Laue's report: *Über die vierzähligen und dreizähligen Photogramme der Zinkblende und das Spektrum der Röntgen-Strahlung.*

390. Letter from E. Solvay to Marie Curie, BNF, Gallica, *Pierre et Marie Curie. Papiers II*, docs.268-269; see note 353.

391. Unlike Lord Kelvin, who had joined Rutherford's opinion in 1904, Solvay didn't share this "generally accepted view'.

392. BNF, Gallica, *Pierre et Marie Curie. Papiers II*, doc.270. The authors thank again Nicolas Coupain for providing them with a copy of this letter.

393. E. Solvay, Energie radioactive de transformation, BNF, Gallica, *Pierre et Marie Curie. Papiers II*, doc. 275.

394. *Ibidem*, doc.276.

395. The minutes of these meetings are kept in notebooks bearing the title "Comité scientifique Solvay", Archives de l'ESPCI, Fonds Langevin, L008, 1-5. These documents show that the granting of subsidies, ISIP's least known activity, was the one which required the greatest efforts. The ISC examined about hundred research projects during the fifteen months between June 1912 and August 1914, and granted 40 Solvay subsidies.

396. Lorentz had to overcome many obstacles, in particular the ban on communicating by letter with his Belgian correspondents: the Germans imposed from June 1915 the exclusive use of postcards.

397. C.G. Darwin, Recent Extensions of the Quantum Hypothesis, *Nature*, May 14, 1914.

398. Arnold Eucken (Ed.), *Die Theorie der Strahlung und der Quanten. Mit einem Anhänge über die Entwicklung der Quantentheorie von Herbst 1911 bis zum Sommer 1913*, Halle, Wilhelm Knapp, 1914.

399. N. Bohr, *Philosophical Magazine* 26, 1, 1913. See A. Pais, 1991, op. cit., 146.

400. Tassel insisted that the text of this conference be published by ISIP after the war (see his letter to Lorentz of February 6, 1919, FIS, S.a.b.ULB, doc.685b). But Lorentz didn't communicate the text, invoking political tensions between Belgium and the Netherlands (see his letter to Tassel of March 7, 1919, *ibidem*, doc.686). It should be noted that an article by Lorentz, entitled "Considérations élémentaires sur le principe de relativité" had appeared in 1914 in the journal *Revue générale des sciences*, 25, 179 (see P. Zeeman and A. D. Fokker, op. cit., vol. 8, 390).

401. This conference was published, see P. Zeeman and A. D. Fokker, op. cit., vol. 8, 390.

402. Letter from Tassel to Lorentz of March 17, 1914, FIS, S.a.b.ULB, doc.1326.

403. Letter from Solvay to Ostwald of November 11, 911, Archiv der Berlin-Brandenburgischen Akademie der Wissenschaften. The authors thank Diana K. Buchwald and Dr. Vera Enke for having provided them with copies of Solvay's letters to Ostwald of July 7 and November 11, 1911.

404. Letter to Solvay from representatives of Chemical Societies, April 13, 1912, FIS, S.a.b.ULB, dossier chimie.

405. Letter of April 19, 1911, already mentioned in section 4.3, note 340.

406. Lafontaine was a Belgian senator, co-founder with Otlet of the International Institute of Bibliography. Committed pacifist, he obtained a Nobel Prize for Peace in 1913.

407. Chemist and pharmacist, Charles Moureu was a relentless researcher, director since 1907 of the journal *La Revue scientifique*. In 1911 he was elected member of the Institut de France.

408. Letter from Lorentz to Solvay of August 17, 1912, FIS, S.a.b.ULB.

409. Letter from Solvay to Lorentz of August 22, 1912, *ibidem*, doc.149.

410. Letter from Van Laer to Solvay, August 1912, FIS, S.a.b.ULB, dossier chimie.

411. Letter of Haller to Tassel of January 2, 1913, FIS, S.a.b.ULB, dossier chimie. The other letters cited in this section are part of the same file (most of these documents have not been subject to date of an inventory).

412. F. Birck, op. cit., 159.

413. Letter from Solvay to Ostwald of January 20, 1913, FIS, S.a.b.ULB, see note 411.

414. *Ibidem*, letter from Tassel to Haller of January 20, 1913.

415. *Ibidem*, letter from Ostwald to Solvay.

416. *Ibidem*, letter from Solvay to Ostwald of January 27, 1913.

417. *Ibidem*, letter from Tassel to Solvay.

418. Solvay had apparently been struck by Berthollet's writings, see Isabelle Stengers, La Pensée d'Ernest Solvay et la science de son temps, in A. Despy-Meyer and D. Devriese (Eds.), op. cit., 152.

419. J. Dhombres and J.B. Robert, *Fourrier créateur de la physique mathématique*, Paris, Belin, 1998, 224.

420. This law was Dewar's guide in 1898 for the liquefaction of hydrogen; in 1908, it guided Kamerlingh Onnes to the liquefaction of helium.

421. Moniteur, cinquième série, t. III, 1913, 126-127.

422. As mentioned earlier, Solvay wouldn't refrain from asking help from well-chosen members of ISIP's Scientific Committee (see our comments of section 4.4).

423. Letter from Haller to Tassel of April 14, 1913, FIS, S.a.b;ULB, dossier chimie.

424. Ostwald used to associate an energetic content with every action.

425. Letter from Guye to Haller of May 7, 1913, Archives de l'ESPCI Paris Tech, Fonds Albin Haller, carton 18. The authors thank Mrs. Catherine Kounelis, director of the Archives, for providing access to this document.

426. Haller wanted to surprise Solvay with the announcement. But the decision had been taken "in secret committee" and a regulation of the Academy provided that the interested party be informed without any delay. Solvay therefore learned about it on August 13, 1913. He immediately expressed his gratitude to Haller.

427. *Proceedings of the Third Session of the Council of the International Association of Chemical Societies*, September 13-23, 1913, 24, Archives of the IACS. These undisclosed documents are now part of the Archives of the Belgian Chemical Society (see Annex 2).

428. The minutes of these meetings are recorded in notebooks entitled "Comité scientifique Solvay. Procès-verbaux et Demandes de subsides". Archives de l'ESPCI, Paris Tech, Fonds Langevin (see fig.26).

429. FIS, S.a.b.ULB, doc.1122

430. Letter from Solvay to Brillouin of March 5, 1913, *ibidem* doc.1123.

431. *Ibidem*, doc.1124.

432. Following the transfer of 6,000 francs taken from the scholarship budget (these were only three in number and represented a total of 11,000 francs), the subsidy budget for the year 1912-1913 amounted to 25,000 francs. Taking into account the 7,500 francs allocated to Laue and Barkla, the ISC realized in June 1913 that it could spend 17,500 francs in subsidies.

433. The following notes are kept in Paris, Archives de l'ESPCI, Fonds Langevin, L10/129.

434. Walther Friedrich and Paul Knipping worked under Sommerfeld's direction. They realized the experiment designed by Laue to detect the X ray deflections produced by a crystal.

435. Letter from Lorentz to Tassel, FIS, S.a.b.ULB, doc. 28.

436. Letter from Solvay to Tassel of July 21, 1913, *ibidem*, doc.1131.

437. In reality, the ISC needed a complement of 7,350 francs: 17,050 francs were needed to honor the requests 1-8; it meant that 450 francs (of the 17,500 francs available) could still be used to honor the requests 9-15.

438. Letter from Tassel to Lorentz of July 23, 1913, FIS, S.a.b.ULB, doc.1132.

439. *Ibidem*, doc.1140.

440. Franck & Hertz: "Ein Teil der verwandte Apparate, so wie das benutzte Platin is aus Mitteln der Solvay Stiftung angeschaffte wofür wir unseren besten Dank auszusprechen haben", *Verhandlungen der deutschen physikalischen Gesellschaft*, 16, 1914, 457.

441. J. L. Heilbron, *Lectures on the History of Atomic Physics 1900-1922*, Proceedings of the International School of Physics, Course LVII, 1972, New York, Academic Press, 1977, 74-75.

442. J. L. Heilbron, *H. G. Moseley: The Life and Letters of an English Physicist*, Berkeley, 1974, 1887-1915.

443. Another beneficiary of a Solvay grant died on the battlefields of the Great War: the Hungarian physicist G. Zemplén.

444. FIS, S.a.b.ULB, doc.1233.

445. Archives de l'ESPCI, Fonds Langevin, doc.L12/83.

446. H. Moseley, The High Frequency Spectrum of the Elements. Part II, *The Philosophical Magazine* 27, 1914, 703-713.

447. J. Stark, *Sitzungsberichte der königlich preussischen Akademie der Wissenschaften*, Sitzung der 20 Nov.1913.

448. Letter from Stark to Lorentz of November 20, 1913, NHA, 364, inv. num.75. NB: some authors wrongly claim the letter got lost.

449. These presentations used to take place in the evening. They started at 8,15 PM, and continued without time limit (with or without a break) according to the speaker's wish. The audience included freshmen, but also top scientists, such as Lorentz, Zeeman, Ehrenfest and Debije.

450. Letter from Lorentz to Tassel, FIS, S.a.b.ULB, doc.1299.

451. Letter from Tassel to Lorentz of February 27, 1914, *ibidem*, doc.1250.

452. *Ibidem*, doc.1252.

453. *Ibidem*, doc.660.

454. Letter from Lorentz to Héger of August 13, 1914, *ibidem*, doc.327.

455. Brillouin referred to the Dreyfus case.

456. Letter from Wien to Lorentz of October 7, 1914, A. J. Kox, 2008, *op. cit.*, 395.

457. Lorentz would answer later that he would rather publish in Dutch journals.

458. Warburg intervened to facilitate the publication of the article. It was important, in his eyes, not to hurt the feelings of the German public, as rumors had circulated on abuses committed by Belgian civilians on wounded Germans (these rumors were later contradicted by testimonials reported by the Dutch Red Cross; Lorentz transmitted them to certain German colleagues, who declared themselves convinced).

459. Letter from Planck to Lorentz of November 28, 1914, A. J. Kox, *op. cit.*, 398.

460. Sommerfeld's attitude had changed during the last months. Thinking, like many others, that the war wouldn't last, he had said to Langevin in a letter of October 19, 1914: "*Let's hope that scientific relationships won't be destroyed this year...*". See, A. Langevin, *op. cit.*, 7.

461. A. Sommerfeld, *Wissenschaftlicher Briefwechsel*, Band I: *1892-1918*, 489 (letter of December 25, 1914).

462. S. L. Wolff, *op. cit.*, 337.

463. Letter from Stark to Lorentz of December 12, 1914, see A. J. Kox, 2008, *op. cit.*, 401. Despite this incident, Lorentz and Stark continued for a year their scientific correspondence.

464. Hermann Kellermann, *Krieg der Geister*, Dresden, 1915.

465. FIS, S.a.b.ULB, doc 836.

466. Elements in support of this claim were presented in a book by Gabriel Petit and Maurice Leudet "Les Allemands et la science" which appeared in 1916. The British chemist Ramsay, and twenty-seven French scholars, used this publication to defend their concept of German science. Ramsay insisted on the limited role played by the Germans in scientific discoveries. The mathematician Picard recommended the suspension, after the war, of all contact with German intellectuals. "*It's essential, he wrote, to free us from the fog of German thought and of the German systematic mind. We will take notice of certain German works, but without the need of personal relations. We hope to be able to organize congresses with our allied friends, from which Germany will be excluded*".

467. Letter from Voigt to Lorentz of February 1915, A. J. Kox, 2008, *op. cit.*, 281.

468. NHA, 364, inv. num.11.

469. See Annex 6, The second Moroccan crisis and the Caillaux affair.

470. We already mentioned the discussions between Brillouin and Wien about the "Agadir crisis" which took place during the Council of October-November 1911.

471. Rutherford referred to the use of gazes, tested by Haber near Ypres for the first time in April 1915 (gazes would also be used by British forces).

472. The disagreement between Wien and Lorentz didn't prevent mutual esteem. In 1918 Wien would reiterate his proposal to award a joint Nobel Prize to Lorentz and Einstein for their works in relativity theory.

473. NHA, 364, inv. num. 11.

474. *Ibidem*, inv. num. 35, 61.

475. Having decided to abandon physics, Hostelet oriented himself towards sociology.

476. Letter of March 24, 1917, sent to Lorentz by the German Imperial Legation in The Hague, NHA, 364, inv. num. 632.

477. We already reported Sommerfeld's reaction to Lorentz's tribute to Solvay, but it merely reflected a personal position with regard to a colleague.

478. *Kriegshefte der Süddeutsche Monatschefte*, April 1918, 44.

479. M. Eckert and Karl Märker, *Arnold Sommerfeld Wissenschaftlicher Briefwechsel*, Band I, Berlin, 2000, 454.

480. M. Eckert, *Arnold Sommerfeld: Science, Life and Turbulent Times, 1861-1951*, Berlin, Springer, 2013, 292.

481. See Lorentz's letter to Lefébure of January 7, 1919, FIS, S.a.b.ULB, doc.670.

482. The authors thank Michael Eckert for having provided them with this information.

483. This was only one of Nernst's activities during the war. For a comprehensive report, see Hans-Georg Bartel and Rudolf Huebener, *Walther Nernst: Pioneer of Physics and Chemistry*, World Scientific, 2017, 236-257.

484. R. Brion and J.-L. Moreau, *op. cit.*, 197-200. See also H.-G. Bartel and R. Huebener, *op. cit.*, 254-256.

485. R. Brion and J.-L. Moreau, *op. cit.*, 198.

486. See Liane Ranieri, *Emile Francqui ou l'intelligence créatrice*, Bruxelles, Editions Duculot, 1985.

487. F. Stockmans, *op. cit.*, 323.

488. NHA, 364, inv. num. 56.

489. FIS, S.a.b.ULB, doc.1941.

490. Dr. Goldschmidt from Brussels, Dr. de Broglie from Paris and Dr. Lindemann from Berlin.

491. *Ibidem*, doc.1944.

492. *Ibidem*, doc.1947.

493. *Ibidem*, doc.1955.

494. *Ibidem*, doc.1956.

495. See Tassel's letter to Brillouin of June 13, 1919, *ibidem*, doc.695b.

496. J. Pelseneer, *op. cit.*, 32.

497. Liane Ranieri, *Danny Heineman. Un destin singulier, 1872-1962*, Bruxelles, Editions Racine, 2005.

498. Solvay's Note of November 15, 915, Archives of the Belgian Chemical Society, undisclosed source (see Annex 2).

499. This French chemist, known for his principle of chemical equilibria, was granted the Davy Medal of the Royal Society in 1916. This son of a leading industrialist was an active member of the "Société d'encouragement pour l'industrie nationale", one of the societies which paid homage to Solvay in September 1913.

500. René Purnal, *Ernest Solvay. Conscience de ce temps*, Edition familiale pour le centenaire de sa naissance, Bruxelles, 1938, 176.

501. Solvay's Note of May 25, 1916, Archives of the Belgian Chemical Society, undisclosed source (see Annex 2).

502. Letter from Solvay to Vandervelde, FIS, S.a.b.ULB, doc.680.

503. Letter from Solvay to Haller of August 13, 1916, FIS, S.a.b.ULB, doc.681.

504. Solvay reiterated his proposal for an "International Consultation" in January 1919, when it came to ISIP's future. See his Note "A propos de la letter de M. Brillouin", *ibidem*, doc. 674.

505. *Ibidem*, doc.678.

506. While defending himself from wanting to influence his partners, Solvay didn't hesitate to reaffirm his confidence in the universal role of physical energetics.

507. The authors are grateful to Mrs. Catherine Kounelis, director of the Archives of the ESPCI Paris Tech, and to Mr. Erik Langlinay from the EHESS, for communicating this document.

508. See Solvay's Note of June 8, 1916: "*It will soon become clear that the "old metaphysico-dynamical physics" must be abandoned*", Archives of the Belgian Chemical Society (see Annex 2).

509. The authors thank Erik Langlinay for communicating this information. See also Brigitte Van Tiggelen and Danielle Fauque, The Formation of the International Association of Chemical Societies, *Chemistry International* 34-1, 2012, and Danielle Fauque, French Chemists and the International Reorganization of Chemistry after WWI, *Ambix* 58-2, 2011, 116-135.

510. A. Haller, President; M. Hanriot, Vice-president; A. Béhal, General Secretary.

511. J. Pelseneer, *op. cit.*, 76.

512. This Union, from which chemists from the central empires were excluded, was created in 1919 by the International Research Council. Germany would be invited to join in 1929 but would again be excluded during the Second World War.

513. See Mary Jo Nye, Chemical Explanation and Physical Dynamics: Two Research Schools at the First Solvay Chemistry Conferences, 1922-1928, *Annals of Science* 46, 1989, 461-480.

514. The signatures of Solvay, Tassel and Warnant appear on the Menu of a dinner which took place on May 19, 1916, at The Hague's Hôtel des Indes; see Archief Kamerlingh Onnes, RMB, Leiden.

515. Letter from Lefébure to Solvay, FIS, S.a.b.ULB, doc.670.

516. Letter from Lorentz to Solvay of January 10, 1919, *ibidem*, doc.683.

517. Solvay's handwritten note on Tassel's letter to Lorentz, NHA, 364, inv. num. 78.

518. FIS, S.a.b.ULB, doc.675.

519. *Ibidem*, doc.671.

520. *Ibidem*, doc.675.

521. Letter from Marie Curie to Lorentz, NHA, 364, inv. num.16.

522. *Ibidem*, inv. num.86 (letter from Warburg to Lorentz of February 23, 1919) and inv. num.56 (letter from Nernst to Lorentz of February 29, 1919).

523. Goldschmidt expressed his displeasure in a letter to Lorentz of March 31, 1920, NHA, 364, inv. num.26.

524. F. Stockmans, *op. cit.* See also https://en.wikipedia.org/wiki/Robert_Goldschmidt.

525. FIS, S.a.b.ULB, doc.695.

526. Let us recall that the Minutes of the first post-war ISC meetings are kept at the Archives of the ESPCI: Paris ESPCI-PSL, Centres de ressources historiques, Fonds Langevin (Fig.26).

527. The average annual salary of a Belgian mason, 1,500 francs in 1910, was about four times higher in 1920 (5,940 francs).

528. See Righi's statement in his letter to Solvay of April 14, 1920 (FIS, S.a.b.ULB, doc.752): "*Unfortunately, the relativists form today an organized party, in which one might perhaps find traces of race or of political profession... it even seems that a conspiracy of silence has been set up around my research*".

529. Lorentz's decision is indicative of his impartiality and high conscience as ISC chairman. Indeed, we know that the precursor of the special theory of relativity was highly interested in the general theory, the success of which had propelled Einstein to the forefront of the international scene.

530. Einstein, officially a civil servant of the German Reich, could still be invited on account of his Swiss citizenship (Lorentz may have known that Einstein had taken care to keep his precious Swiss passport).

531. The humiliation of German physicists was such that in 1926 they refused to join the International Union of Pure and Applied Physics (IUPAP) despite the lifting of their exclusion. Germany wouldn't join before 1954 (Austria waited until 1957).

532. *Atomes et Electrons. Rapports et discussions du troisième Conseil de physique tenu à Bruxelles du 1er au 6 avril 1921 sous les auspices de l'Institut international de physique Solvay*, Paris, Gauthier-Villars, 1923.

533. It was during the ISC meeting of April 6, 1921, that Langevin was asked to take the seat of Righi (who had died on June 8, 1920).

534. Einstein had been asked to join he zionist leader Chaim Weizmann in order to raise funds for the creation of Jerusalem's Hebraic University.

535. Bragg senior, member of the ISC, hadn't been able to take part in the Council.

536. Michelson, who was in London at the time, was invited "to replace" Einstein.

537. See Tassel's letter to Brillouin of February 23, 1921, FIS, S.a.b.ULB, doc.2066.

538. Lindemann had argued that Cambridge was represented by several members, and that he was the only physicist who could represent Oxford. He should have known that this argument couldn't be accepted, the members being invited in a personal capacity. In fact, only two invitees were from Cambridge: Rutherford and Sir Joseph Larmor.

539. Brillouin considered Lindemann as a "spy" sent by Nernst (see Brillouin's letter to Tassel of February 13, 1921, FIS, S.a.b.ULB, doc.2065.

540. De Broglie made a decisive remark: "*An atom which receives light of frequency v emits a projectile, endowed with energy hv, long before the radiation has been able to provide it with the elements of this energy by means of homogeneous spherical waves*".

541. See P. Langevin and M. de Broglie, *op. cit.*, 454.

542. Siegbahn would be awarded a Nobel Prize in 1924 for his work on spectroscopy and X rays.

543. See A. Pais, 1991, *op. cit.*, 214.

544. Adèle Solvay's telegram to Lorentz, NHA, 364, inv. num.73.

545. See Armand Solvay's speech, FIS, S.a.b.ULB, doc.2407.

546. *Conductibilité électrique des métaux et problèmes connexes. Rapports et discussions du quatrième Conseil de physique tenu à Bruxelles du 24 au 28 avril 1924 sous les auspices de l'Institut international de physique Solvay*, Paris, Gauthier-Villars, 1927.

547. Two "outgoing" ISC members were replaced during the Committee meeting: M. Brillouin and E. Rutherford gave up their chair to Prof. Ch.-E. Guye from the University of Geneva, and to Prof. O. W. Richardson from London University.

548. Willem Keesom replaced Heike Kamerlingh Onnes, who had fallen ill.

549. Two leading American physicists had already been invited to Solvay III.

550. Schrödinger would become famous in 1926 for having laid the foundations of *wave mechanics*, a new formalism based on Louis de Broglie's revolutionary idea of "matter waves".

551. We may assume that Joffé had been invited at the suggestion of Langevin, who had joined the ISC and wished to support Soviet physics (see A. Langevin, *op. cit.*, 10-12).

552. The public had clearly been struck by the amputations performed on renowned radiologists.

553. Were present at the meeting: Lorentz, Knudsen, Langevin, Ch.-E. Guye, Richardson (Marie Curie and E. van Aubel were excused).

554. A. Langevin, *op. cit.*, 11.

555. See Bordet's telegram to Lorentz of April 2, 1926 (Jules Bordet was King Albert's ISC representative).

556. See Lefébure's letter to the King of March 25, 1926, FIS. S.a.b.ULB, doc.826.

557. See Lefébure's note of April 16, 1926, *ibidem*, doc.2573.

558. Reminder: Léon Brillouin was Marcel Brillouin's son.

559. In 1924 Schrödinger was working in Zurich, and therefore listed as a Swiss member. In 1926 he was presented as an Austrian member, presumably to underline the presence of representatives of the former central empires.

560. Contrary to Planck, Sommerfeld didn't belong to the list of signatories of the "Manifesto of the 93".

561. Letter from Lorentz to Sommerfeld of November 23, 1924, NHA, 364, inv. num.74.

562. Lorentz was the man who on May 20, 1922, signed Sommerfeld's nomination as a foreign member of the Koninklijke Hollandsche Maatschappij der Wetenschappen.

563. Letter from Born to Sommerfeld, see M. Eckert and K. Märker, *Arnold Sommerfeld Wissenschaftlicher Briefwechsel*, Band 2, Diepholz-München, 2004.

564. The details of the Council and its preparation have been reported by Guido Bacciagaluppi and Antony Valentini in *Quantum Theory at the Crossroads: Reconsidering the 1927 Solvay Conference*, Cambridge University Press, 2009. We therefore focused on some further elements, taken from other works, such as Patrick Coffey's book: *Cathedrals of Science. The Personalities and Rivalries that Made Modern Chemistry*, Oxford University Press, 2008, and from documents which belong to the International Solvay Institutes (S.a.b. ULB).

565. Letter from Lorentz to Lefébure of April 27, 1927, FIS, S.a.b.ULB, doc.2574.

566. Einstein also wished to remain discrete about his (unpublished) results regarding the ideas of L. de Broglie, see G. Bacciagaluppi and A. Valentini, *op. cit.*, 13.

567. See Ehrenfest's letter to Lorentz of March 30, 1926, NHA, 364, inv. num.20.

568. FIS, S.a.b.ULB, doc.2523.

569. Lefébure's letter to Lorentz of October 14, 1927, *ibidem*, doc.2534.

570. *Ibidem*, doc.2536.

571. This Conference organized in Como to commemorate the centenary of Alessandro Volta's death, shortly preceded Solvay V.

572. See Lorentz's letter to Lefébure of October 9, 1927, FIS, S.a.b.ULB, doc.2548.

573. *Ibidem*, doc.2541.

574. See P. Coffey, *op. cit.*, 188.

575. *Ibidem*, 188-190.

576. Langmuir in his letter to his mother, see the previous note.

577. *Ibidem*, 190.

578. See Youtube, FreeScienceLectures.com, The 1927 Solvay Conference.

579. See M. Brillouin's letter to Lorentz of october 11, 1927, NHA, 364, inv. num.10.

580. Letter from L. Lumière to Lorentz of October 11, 1927, FIS, S.a.b.ULB, doc.2615.

581. Page 2 of Brillouin's letter to Lorentz of October 11, 1927, see note 578.

582. See Langevin's speech, FIS, S.a.b.ULB, doc.4601c.

583. Letter from Lefébure to Lorentz of November 10, 1927, *ibidem*, doc.2602.

584. *Electrons et Photons. Rapports et discussions du cinquième Conseil de physique tenu à Bruxelles du 24 au 29 octobre 1927 sous les auspices de l'Institut international de physique Solvay*, Paris, Gauthier-Villars, 1928. More details can be found in G. Bacciagaluppi and A. Valentini, *op. cit.*, 18.

585. According to this principle the apparently contradictory behaviours of light and electrons should be seen as a manifestation of two "mutually excluding" aspects of a single physical reality, determined by the observational conditions.

586. This effect results in a variation in the wavelength of X-rays, following their scattering by electrons. Compton shared the Physics Nobel Prize of 1927 with C. T. R. Wilson, other Council member and inventor of the "cloud chamber" (also called Wilson chamber).

587. Elsasser had no luck. His conclusions, published in *Naturwissenschaften* and mentioned by Born in 1926, weren't reported by Davisson and Germer in their paper of 1927. His absence from Solvay V reminds us of Ehrenfest's (more problematic) absence from Solvay I.

588. W. Heisenberg, Theory, Criticism and a Philosophy, in *From a Life of Physics*, World Scientific, 1989, 47.

589. A point emphasized by G. Bacciagaluppi and A. Valentini in their book, *op. cit.*, 242-245.

590. *Ibidem*, 19 and 244.

591. N. Bohr, 1962, *op. cit.*, 26-28.

592. *Ibidem*, 30. Einstein had relied on the equivalence between mass and energy, consequence of the special relativity theory. Bohr pointed out that he had not taken into account a consequence of the general theory: the effect of a gravitational field on the operation of a clock.

593. G. Lemaître, Rencontres avec A. Einstein, *Revue des questions scientifiques, 20 janvier 1958* (the text was read by Lemaître on Belgian national radio on April 27, 1957, on the occasion of the second anniversary of Einstein's death).

594. Caesar Koch, Einstein's favourite uncle (brother of his mother and father of his second wife Elsa) was a Swiss citizen who lived since many years in Belgium. He stayed, at the time of the Council, with his daughter Suzanne Koch-Gottschalk. The authors are grateful to the members of the Ferrard family, direct descendants of Einstein's Belgian cousin.

595. W. Heisenberg, Die Entwicklung der Quantentheorie 1918-1928, *Die Naturwissenschaften*, 14, 1956, 490-496.

596. Solvay repeatedly provided financial support to the Leiden cryogenics laboratory, and visited it three times during the war (Dirk van Delft, *Freezing Physics: Heike Kamerlingh Onnes and the Quest for Cold*, Amsterdam, Edita, 2007)

597. H. Kamerlingh Onnes, *op. cit.*, 554.

598. Draft of Lorentz's letter to Lefébure of September 29, 1922, and intended for the Solvay family, NHA, 364, inv. num.73. The authors thank Mrs. Marina Solvay for providing them with elements which enabled them to establish the link between the above document and Lefébure's confidential letter.

599. H. Kamerlingh Onnes, *op. cit.*, 555.

600. G. Doyer van Cleeff, *Scheikunde in dienst van den Mensch*, Haarlem, H. D. Tjeenk Willink & Zoon, 1918 (the book contains a portret of Ernest Solvay).

601. J. Dhombres and J.-B. Robert, *op. cit.*, 365.

602. J. J. Heirwegh and M. Peeters, *op. cit.*, 168.

603. See Helm's criticism of W. Thomson and P. G. Tait's "Treatise on Natural Philosophy", in *Die Energetik nach ihrer geschichtlichen Entwicklung*, Leipzig, Velt & Comp., 1898.

604. See Solvay's Note of September 15, 1910, "Science et Univers objectifs", Archives of the Belgian Chemical Society (see Annex 2).

605. Letter from Solvay to Berthelot's son of May 7, 1907, Archives of Collège de France. The authors are grateful to Mrs. Anne Chatellier for providing them with a copy of the letter.

606. Kelvin associates matter with '*places where the ether is animated by whirlwind movements*".

607. H. Poincaré, 1905, *op. cit.*, 195.

608. K. Pearson, *The Grammar of Science*, London, A. and C. Black, 1900, 166.

609. A translation in English of Einstein's tribute to Planck, entitled "Max Planck as a Scientist" has been reported in CPAE, vol. 4, doc.23.

610. See E. Crawford, 1987, *op. cit.*, 135.

611. Planck refers to the fact that the value of Avogadro's number can be deduced from thermal radiation data.

612. Planck recalls that the electric charge of the electron (which can be deduced from Avogadro's number), conforms to what Rutherford has obtained from observations linked to radioactivity.

613. The elements that we report here have been borrowed from two sources: Jean-Claude Allain, "Agadir 1911. Une crise impérialiste en Europe pour la conquête du Maroc, Université Paris I, 1976, and Laurence van Ypersele, "Le Roi Albert. Histoire d'un mythe", Quorum, 1995.

614. The case had significant consequences. Aware of the weakness of its encryption system, Germany decided to change it. France was therefore unable to decipher the new German messages before the outbreak of the First World War.

615. Letter from Solvay to Jungbluth, FIS, S.a.b.ULB, doc.151.

616. *Ibidem*, doc.158.

617. Solvay knew that the appointment by the King of a representative in ISIP's Administrative Commission could set an embarrassing precedent. He therefore made sure that the request be perceived as emanating from an International Committee.

618. FIS, S.a.b.ULB, doc.165.

619. Thomson recalled that the helium atom was supposed to be an ordinary constituent of more massive atoms (i.e., atoms belonging to a set of elements much larger than that which comprised radioactive elements).

620. Thomson was prepared to admit the existence inside the atom of forces other than those of ordinary electrostatics. He had shown in 1910 that Planck's radiation law could be obtained from classical dynamics upon the introduction of a law of repulsion in inverse ratio to the cube of the distance.

621. Thomson's idea was a striking foreshadowing of the forces acting between the nucleons (protons and neutrons) that make up the atomic nucleus.

Index

A

Afanassieva, T. 104, 288
Arrhenius, S. 35, 38–39, 85, 178, 194, 245, 276
Asquith, H. H. 284
Aubel, E. van 214, 216, 219, 223, 228, 301

B

Bacciagaluppi, G. 265, 301–302
Barkla, C. G. 4, 136, 167–168, 176 216, 219, 222, 245, 295
Barlow, W. 137, 139, 146
Bartel, H.-G. 265, 298
Bauer, H. 223
Beatty, R. T. 169
Becquerel, A. H. 84, 276
Bédier, J. 195
Beller, M. 266, 268, 274
Berends, F. 265, 267–268, 274
Berthelot, M. 202, 250, 303
Berthollet, C.-L. 157, 294
Besso, M. 3, 34, 80, 97, 274, 285–287, 289
Bestelmeyer, A. 169
Birck, F. 265, 283, 294
Bohr, N. 82, 132, 146, 176–177, 197, 216, 219–223, 227, 231–232, 234–237, 245, 265, 267, 283–284, 286, 292, 294, 303
Bolle, J. 275, 287
Boltzmann, L. 14, 33, 38, 44, 56, 65, 70, 102, 104, 253, 277, 280
Bonaparte, R. 59
Borel, E. 84–85

Born, M. 40, 47, 83, 146, 227, 229, 231–236, 238, 245, 301–302
Bose, S. N. 228
Boutaric, A. 22, 265, 276
Bragg, W. H. 136–137, 139, 141–142, 146, 176, 191, 194, 213, 215–216, 221–223, 228, 232, 245, 292–293, 300
Bragg, W. L. 4, 141, 170, 176, 180, 194, 216, 219–221, 227, 234–235, 238, 245, 293
Bridgman, P. W. 223–226, 245
Brillouin, L. 216, 219, 221, 223, 227, 230–232, 234, 280, 301
Brillouin, M. 2, 14, 16, 50–52, 73–75, 81–82, 86, 88, 95, 116–117, 119, 123, 134, 137, 139, 145, 171, 173, 181, 185–186, 192, 195, 204, 212, 214–217, 219, 221–223, 281, 285–287, 290–291, 295–298, 300–302
Brion, R. 197, 265, 279, 298
Broglie, L. de 82, 227, 231–236, 245, 286, 301–302
Broglie, M. de 14, 16, 55, 70, 72–73, 76, 137, 139, 145, 216, 219–220, 222, 227, 265–266, 268, 273, 280, 284–287, 298, 300
Broniewski, W. 223–224
Bustamante, M. C 281

C

Caillaux, J., 192, 257–258, 283, 285, 297
Carnegie, A. 17, 156, 161, 189
Chadwick, J. C. 222, 245
Clark, R. 265, 287, 289–290

Clausius, R. 14, 44, 56, 65, 70
Coffey, P. 230, 265, 301–302
Cohen, R. 266, 268, 274
Compton, A. H. 30, 225, 227, 232, 234–236, 245, 302
Cotton, A. E. 216
Coupain, N. 247, 265, 275–276, 287, 291, 293
Couprie, B. 15
Crawford, E. 265, 273, 278–279, 286, 304
Crick, F. 293
Curie, M. 2, 4, 14–16, 39, 50, 55, 73, 75, 79–80, 82–88, 95, 98–99, 107, 119, 123, 126, 129, 134, 137, 144–145, 180–181, 213, 215–216, 219, 223, 226, 234, 238–239, 245, 266–267, 279, 282, 286–287, 291–293, 299, 301
Curie, P. 84, 87, 267, 293

D

Danysz, J. 170
Darwin, C. G. 146, 193, 227, 279, 293
Davis, B. 227
Davisson, C. T. 235, 302
Debije (ou Debye), P. 97, 100–101, 103, 143, 146, 223, 227, 231–232, 234, 245, 288, 296
Dechend, E. von 168, 174
De Greef, G. 19
Delft, D. van 303
Dember, H. 168, 174
Denis, H. 19
Deslandres, H. A. 227–228
Despy-Meyer, A. 247, 266, 268, 275, 294
Devriese, D. 247, 266, 268, 275, 294
Dewar, J. 157–158, 267, 276, 295
Dhombres, J. 265, 295, 303
Dirac, P. A. M. 227, 229, 231, 233–236, 245
Dobronravoff 224
Donnan, F. 60, 62, 283
Dony-Hénault, O. 60
D'Or, L. 247, 265, 275
Dreyfus, A. 296

Dulong, P. L. 13, 21, 31–32, 43, 46, 49, 127
Dunoyer, L. 169

E

Eckert, M. 196, 265, 267–268, 274, 285, 297–298, 301
Ehrenfest, P. 28, 103–105, 122, 131, 179, 216, 219, 221, 227–228, 231, 234, 236, 266, 277, 288–289, 296, 302
Einstein, A. 2–4, 7–9, 13–16, 23, 28–35, 37, 41, 43, 49–50, 52–56, 65–67, 70–75, 78–82, 85–86, 95–102, 104–105, 107–111, 136–137, 141–143, 146, 191, 211, 215–216, 220–223, 225–228, 231–232, 234–237, 245, 253, 265–266, 268, 274, 277–282, 284–290, 292–293, 297, 300, 302–303
Elsasser, W. 235, 302
Enke, V. 294
Errera, J. 229
Eucken, A. 146, 268, 278, 285, 294
Eve, A. 266, 279, 286

F

Fabry, Ch. 118, 216, 227, 290
Fajans, K. 169
Faraday, M. 147, 287
Fauque, D. 299
Fischer, E. 25, 39, 49, 282
Fokker, A. D. 195, 267, 277, 289, 293–294
Forrer, L. 96
Fournier d'Albe, E. E. 169
Fowler, R. H. 227, 232, 234
Franck, J. 4, 169, 172, 174, 176, 245, 296
Francqui, E. 198, 201, 298
Fresnel, A. 20, 231, 275
Freundlich, E. 108, 169, 174
Friedrich, W. 38, 108, 168, 174, 295

G

Germer, L. 235, 302
Giroud, F. 266, 286–287
Gockel, A. 169, 173–174

Index

Goldschmidt, R. 3, 14, 16, 23, 40–43, 45–46, 50–52, 55–57, 64, 66, 68 73, 75, 83, 129–130, 134, 137, 146, 174, 197–199, 214, 259, 267, 276, 279, 281, 283–285, 293, 298–299
Grignard, V. 245, 290
Gross, D. 266, 268, 274
Grossmann, M. 97
Grummach, L. 169
Grundmann, S. 266, 277, 289
Grüneisen, E. 136–137, 139, 141, 293
Gubin, E. 83, 266, 286

H

Haas, A. E. 82, 286
Haas, G. de 74, 113
Haas, W. de 216, 219–221, 285
Haber, F. 108–110, 197, 289, 297
Hall, E. H. 189, 223–226
Haller, A. 8, 59, 85, 154–160, 162–166, 203–206, 209–210, 282–283, 294–295, 298–299
Hasenöhrl, F. 14, 16, 43, 52, 99, 102, 107, 137, 193, 279
Hayez (imprimerie) 146, 199–201, 282
Héger, P. 18, 59, 90, 115, 120, 125, 129, 131, 134, 185, 190, 212, 217, 260, 276, 290–291, 296
Heilbron, J. L. 266, 274, 284, 286, 296
Heineman, D. 201, 298
Heirwegh, J.-J. 266, 275, 303
Heisenberg, W. 83, 227, 229, 231–236, 238, 245, 302–303
Henneaux, M. 266, 268, 274
Henriot, E. 216
Hertz, G. 4, 169, 172, 174, 176, 245, 296
Hertz, H. 121, 168, 250, 277
Herzen, E. 3, 16, 21–22, 24, 47–48, 51–55, 57–58, 65–66, 89, 116, 144, 151–152, 201–202, 240, 276, 280–281, 283
Hevesy, G. de 223, 225–226, 245
Hoff, J. van't 110, 282, 289
Hoover, H. 198
Hopf, L. 136, 141–142, 289, 292–293
Hostelet, G. 3, 16, 21, 24, 47–48, 57, 89, 115–116, 151–152, 196, 281, 290, 297

Huebener, R. P. 265, 298
Hupka, E. 168, 173–174

I

Icole, P. 168, 174

J

Jammer, M. 266, 274, 278
Jeans, J. 14, 16, 26, 35, 43, 52, 54, 56, 65, 68, 70, 72, 76, 82, 95, 137, 216, 219, 277, 279, 286, 292
Joffé, A. 223–224, 226, 301
Jorissen, W. T. 169, 174
Julius, W. 20, 96–97, 99–100, 168, 287–288
Jungbluth, H. 130, 259–260, 304

K

Kamerlingh Onnes, H. 3, 14, 16, 41, 55, 71–75, 83, 95–96, 113, 123, 131, 134, 137, 145, 196, 202, 209, 216, 219, 221, 223, 225–226, 235, 238–240, 245, 266, 276, 282–284, 291, 295, 299, 301, 303
Kapitza, P. 227
Kàrman, Th. von 146
Kelvin, Lord 250, 282, 284, 293, 303
Kepler, J. 23, 151
King Albert 19, 56, 64, 123, 128, 130–131, 134, 163, 197–198, 209, 226–227, 257, 259–260, 273, 301, 304
Klein, M. 266, 274, 277–278, 282, 284, 288
Knipping, P. 295
Knudsen, M. 3, 14–16, 41, 43, 52, 55, 65, 71, 73, 123, 128–129, 131–135, 137, 145, 168, 181, 216, 219, 223, 234, 279, 282, 284, 291–292, 301
Koch, C. 237, 303
Koenigsberger, J. 169
Kohlrausch, F. 33
Kohnstamm, Ph. 53
Koppel, L. 109–110, 289
Kormos Barkan, D. 266, 274, 278–280, 286
Kox, A J. 266, 268, 278, 281, 288–289, 296–297

Kramers, H. 227, 231–232, 234
Kuhn, Th. 266, 277–278, 281

L

Laer, H. Van 151–153, 294
Lafontaine, H. 151, 294
Lambert, F. 265, 267–268, 274, 276, 284
Langevin, A. 266, 273, 296, 301
Langevin, P. 2, 14, 16, 43, 52, 71–72, 80, 83–88, 95, 119, 137, 145, 216, 219, 223, 226, 228, 232, 234, 238, 266, 268, 273, 279–281, 283–287, 296, 300–302
Langmuir, E. 229–231, 245, 302
Larmor, J. 14, 41, 43, 52, 68, 216, 219, 279, 300
Laub, J. 33–34, 278
Laue, M. von 4, 135–137, 139, 141–143, 146, 167–168, 170, 176, 180, 214, 245, 292–293, 295
Lazarew, P. 133, 167, 289
Lebedew, P. N. 4, 103–104, 121, 131–134, 167–168, 285, 288–289, 291
Leblanc, N. 17, 164
Lefébure, Ch. 63, 125, 147, 207–208, 228–230, 233, 240, 283, 297, 299, 301–303
Lemaître, G. 237, 273, 303
Lenard, Ph. 136–137, 277, 289
Leudet, M. 297
Levi-Civita, T. 215
Levie, F. 266, 279, 282
Lewis, G. 230
Lindemann, F. 3, 16, 37, 50, 68, 70, 78, 84, 107, 110–111, 137, 139, 145, 194, 221, 223, 229, 274, 279, 284, 298, 300
Lorentz, H. A. 2, 4, 6–9, 14–17, 22, 27–28, 30, 35, 41, 43, 45, 52–55, 63, 65–80, 82, 86, 88, 90, 95–97, 99–107, 113–117, 119–123, 125, 127–131, 133–137, 139, 142, 145, 147, 149–150, 152, 158, 164, 167–168, 170–175, 179–181, 185–188, 190, 192–193, 195–197, 199–205, 208–209, 212–217, 219–221, 223–236, 238, 240, 245, 259, 275–280, 282, 285–297, 299–303
Lorenz, R. 47, 281

Löwenthal, E. 108, 289
Lowry, J. M. 169, 173–174

M

Macé de Lépinay, J. 118, 290
Maecenas, G. C. 17
Manneback, Ch. 229
Marage, P. 266, 268, 273–274, 286
Maric, M. 108
Mascart, E. 87
Massart, J. 211
Maxwell, J. Clerk 4, 14, 27, 34, 44, 56, 65, 70, 79, 95, 121, 277
Mayer, J. 20, 275
Meyer, E. 168, 173–174
Meyer, J. 168, 174
Meyer, S. 85, 286
Millikan, R. 30, 106, 216, 219, 221, 277
Millochau, G. 168, 174
Mittag-Leffler, G. 35, 85
Moissan, H. 118
Mond, L. 24, 276
Moreau, J. L. 197, 265, 279, 298
Muspratt, E. K. 60–61, 283

N

Nagaoka, H. 216
Natanson, W. 185
Nernst, W. 2–3, 5–8, 14–16, 25, 31–35, 37–58, 63–66, 68, 70–73, 78, 83, 93, 95, 97, 107–108, 110, 117, 123, 133–135, 137, 141–142, 145, 158, 174, 181, 188, 195, 197–199, 211, 213–214, 222, 238, 245, 250, 265–267, 275–276, 278–285, 287, 289, 291–292, 298–300
Nicolai, G. 191

O

Oppenheimer, R. 238, 273
Ornstein, L. 53
Oseen, C. 223
Ostwald, W. 5, 7, 19–20, 25, 39, 51, 58–60, 62–63, 114, 149–151, 153–160, 162, 188, 205, 250, 267, 275–276, 279, 282–283, 290, 294–295
Otlet, P. 40, 59, 294

Index

P

Pais, A. 267, 278, 286, 292, 294, 300
Parijs, P. van 144
Pauli, W. 227, 229, 231–232, 234–235, 238, 245, 280
Peeters, M. 266, 275, 303
Pelseneer, J. 267, 280, 298–299
Perrin, J. 2, 14–16, 43, 52, 55–56, 68, 70–72, 80, 84–87, 95, 216, 219, 245, 279, 281, 283
Philippson, F. 197–198, 265, 279
Picard, E. 24, 276, 297
Piccard, A. 228–229, 237
Planck, M. 2, 4, 13–16, 26–35, 37, 41–45, 49–56, 64–67, 70–71, 73, 78–79, 81–83, 104, 108–110, 135, 187–188, 195–198, 208, 211, 227–228, 233–235, 245, 251, 253, 255, 266, 273–274, 277–280, 282–286, 288, 296, 301, 303–304
Poincaré, H. 2, 14–16, 20, 50, 55, 57, 70, 76, 78–79, 81–82, 95, 98–99, 104, 134, 146, 151–152, 250–251, 267, 275, 277, 285–286, 288, 305
Poincaré, R. 87–88, 258
Pope, W. J. 137, 139, 146, 206

R

Ramsay, W. 45, 59, 117, 150–151, 153–159, 162, 164, 191, 203, 285, 297
Ranieri, L. 298
Rayleigh, Lord 14–15, 17, 26, 35, 41, 43–45, 52–55, 65, 68, 70, 117, 191, 245, 275–277, 279, 284
Renn, J. 266, 268, 274
Richardson, O.W. 216, 219–220, 223–224, 232, 234, 245, 301
Righi, A. 121, 213, 215–216, 300
Rockefeller, J. D. 17
Rohn, W. 169
Röntgen, W. C. 45, 52, 55, 79–80, 128, 132, 135–136, 139, 142, 168–170, 261, 279, 292–293
Rosenfeld, L. 267, 274, 277
Rosenhain, W. 223–224
Rowlinson, J. S. 267, 276
Rubens, H. 14, 16, 33, 55, 71, 95, 107, 110, 137, 281, 283–284, 289
Rutherford, E. 3–4, 14–16, 39, 43, 45, 52, 68, 73, 75–76, 79–80, 82, 85, 95, 117, 123, 131–134, 136–137, 140–141, 145–146, 173, 177, 193, 213, 215–216, 219–220, 222–223, 245, 255, 261–262, 266, 279–280, 285–286, 291–293, 297, 300–301, 304

S

Sabatier, P. 118, 290
Salmon, Ch. 169
Schirrmacher, A. 267, 274, 280
Schmidt, G. C. 169, 174
Schmidt-Ott, F. 108
Schollaert, F. 64
Schrödinger, E. 83, 223, 225–227, 231, 233–236, 238, 245, 301
Schuster, A. 14, 33–35, 41, 43, 52, 68, 279, 282
Seeliger, H. von 41, 45, 52, 55, 279
Seeliger, R. 170
Sevrin, A. 266, 268, 274
Siegbahn, M. 216, 219, 222, 245, 300
Siegel, D. 266, 268, 281
Solvay, Adèle 223, 300
Solvay, Armand 223, 300
Solvay, Ernest 1–4, 7, 14, 16–17, 22, 24–25, 39, 48, 52, 59, 89–90, 93, 116–117, 120, 122–131, 134, 144–145, 149–157, 160–166, 171, 174, 179, 185, 187, 201–209, 212–213, 215, 240–242, 247–251, 259, 265–268, 271, 273–276, 279, 281, 283, 285, 287, 290–291, 293–295, 298–300, 303
Sommerfeld, A. 3, 8, 14, 16, 34, 54–56, 65, 71–72, 74, 79, 81–82, 90, 95, 103–105, 135, 137, 143, 146, 168, 174, 187–188, 196–197, 227–228, 239, 278, 284–285, 288–289, 292–293, 295–297, 301
Stark, J. 4, 34, 170, 179–181, 188, 245, 278, 296–297
Stas, J. S. 19, 90–91, 275
Stengers, I. 247, 294

Stockmans, F. 267, 276, 279, 293, 298–299

T

Tassel, E. 20–21, 24, 57, 89–90, 125, 130, 134, 136–137, 147, 152–157, 159–160, 162–163, 165–166, 171, 174–175, 179–180, 199–202, 206, 209, 212–214, 247–248, 260, 267, 274–276, 285, 294–296, 298–300
Téry, G. 85
Thibaut, J. 227
Thiesen, M. 32, 278
Thirring, H. 227
Thomsen, H. P. 202
Thomson, J. J. 14–15, 43, 45, 52, 68, 117, 136–137, 139–141, 146, 191, 194, 199, 216, 222, 245, 261–262, 279, 286, 292, 303–304
Tiggelen, B. Van 287, 299
Trautz, M. 169

V

Vandervelde, E. 19, 189, 203, 298
Verschaffelt, J. E. 75, 130–131, 145, 194, 209, 214, 228, 232, 234, 259
Villard, P. 118, 290
Voigt, W. 105, 136–137, 139, 143, 192, 282, 297

W

Waals, J. D. van der 14–15, 41, 43, 52–55, 68, 83, 101, 157, 245, 279, 284
Wallenborn, G. 266, 268, 273–274, 286
Warburg, E. 14, 16, 55, 70–71, 73, 79–80, 95, 99, 107, 109–110, 117, 123, 134, 137, 145, 174, 181, 187, 196, 199, 208, 211, 213, 283–284, 291, 296, 299
Warnant, E. 202, 209, 299
Watson, J. 293
Waxweiler, E. 19
Weiss, P. 96–98, 136–137, 216, 219, 221, 287–288
Weizmann, Ch. 300
Wien, W. 3, 14–16, 26, 29, 35, 41, 52, 75, 80, 99, 137, 170, 180, 186–188, 193, 195, 245, 253, 279, 286, 296–297
Wilson, C. T. R. 227, 232, 234, 245, 261–262, 302
Wind, C. 100
Wirtz-Cordier, A.-M. 247, 265, 275
Wood, R. W. 137, 139, 170, 292

Z

Zangger, H. 81, 85, 95–98, 286–289
Zeeman, P. 55, 67, 179, 192, 216, 219, 232, 245, 267, 277, 289, 293–294, 296
Zemplén, G. 170, 296

www.ingramcontent.com/pod-product-compliance
Ingram Content Group UK Ltd.
Pitfield, Milton Keynes, MK11 3LW, UK
UKHW021329180426
11947UKWH00017B/1520